PROGRESSIVE KAIZEN

The Key to Gaining a
Global Competitive Advantage

PROGRESSIVE KAIZEN

The Key to Gaining a Global Competitive Advantage

JOHN W. DAVIS

 CRC Press
Taylor & Francis Group
Boca Raton London New York

CRC Press is an imprint of the
Taylor & Francis Group, an **informa** business

A PRODUCTIVITY PRESS BOOK

Productivity Press
Taylor & Francis Group
270 Madison Avenue
New York, NY 10016

© 2011 by Taylor and Francis Group, LLC
Productivity Press is an imprint of Taylor & Francis Group, an Informa business

No claim to original U.S. Government works

Printed in the United States of America on acid-free paper
10 9 8 7 6 5 4 3 2 1

International Standard Book Number: 978-1-4398-4608-7 (Paperback)

Visit the Taylor & Francis Web site at
http://www.taylorandfrancis.com

and the Productivity Press Web site at
http://www.productivitypress.com

Contents

List of Figures xi

List of Tables xiii

Preface.. xv

Acknowledgments xxv

Introduction xxvii

1 Examining the Basics of an Effective Kaizen Process.................1

Matter of "Misguided Pragmatism" 15

Two Major Do's and Don'ts of Kaizen 16

Evaluating and Rating a Company's Kaizen Efforts......................... 17

Developing a Formal Schedule for Kaizen 19

Assigning a Qualified Full-Time Lean/Kaizen Coordinator............ 20

Establishing a Formal Budget for Kaizen 21

Number and Type of Kaizen Events Conducted 22

Scope of Kaizen Training 23

Overview of Various Types of Kaizen 23

Progressive Kaizen Initiative... 28

Precisely What the Term "Event" Means 28

Purpose and Scope of Progressive Kaizen Effort 30

Ensuring Planned Changes Are Carried Out to the Fullest................. 30

Production Manager's Role in a Kaizen Event............................. 32

Key Summary Points.. 33

Elevating the Use and Effectiveness of Kaizen 33

The Four Types of Progressive Kaizen.............................. 33

Developing a Master Plan for Kaizen 33

Developing a Formal Budget for Kaizen............................. 34

Applied Purpose of a Kaizen Event 34

Progressive Kaizen Initiative..................................... 34

2 Addressing Key Roles and Supporting Tactics 35
   Clearing the Five-Inch Hazard 35
   Taking a Close Look at the Distribution of Change 36
   Plant Manager's Role in Lean 38
   Characteristics of Lean-Oriented Plant Managers 41
   Lean Coordinator 43
   Maintenance Manager 44
   F Alliance 44
   Lean-Oriented Company President 45
   Shop Floor Supervisor's Role in Kaizen 48
   Special Consideration of "Owner-Operators" 52
   Value of Inserting a WRAP Initiative 53
   Tactics for Getting the Best Results Out of Kaizen 56
      How to Use the Step Charts 57
   Key Summary Points 60

3 Avoiding the Typical Pitfalls 63
   Allowing Outside Assistance to Cloud a Path to Success 64
   Misstep of Excluding Office Functions 67
      A Special Word about Lean and a Company's Financial Arm 69
   Allowing Kaizen Accomplishments to Deteriorate 70
   Failure to Communicate the Full Extent and Scope of Kaizen 74
   Failure to Effectively Utilize the Production Engineering Function 76
   Failure to Restructure the Stated Objectives of Key Players 79
   Production Manager's Stated Objectives 80
   Shop Floor Supervisor's Stated Objectives 82
   Production Engineer's Stated Objectives 83
   Error of Putting Lean in a Stand-By Mode 84
      First 84
      Second 85
   The Do's and Don'ts Associated with Kaizen 86
      Definite Do's 86
      Definite Don'ts 87
      Preferable Do's 87
      Preferable Don'ts 88
   Simple Exercise for Getting the Most Out of Any Kaizen Effort 88
   Key Summary Points 89
      Staying Focused 89
      Avoiding Slippage 90

Putting Lean Duties in Writing 90
Various Versions of a Process 90
Removing Problems and Enhancing Individual Performance 90
Vital Role of Production Engineering 91

4 Where to Start and How to Proceed 93
Thinking Outside the Box 93
Sum of the Added Cost and Payback of Lean 94
Progressive Kaizen Tool Box 96
Advantages of Labeling Kaizen Activity "Waste Reduction" 100
Value of Putting the First "Pull Zone" in Final Assembly 103
Sticking to the Plan and Avoiding Disruptions 106
Conducting the Factory's First High-Impact Kaizen Event 107
    Basic Event Objectives 108
    Participation of the Production Manager 108
    Participation of Shop Floor Supervisors 111
    Participation of a Key Maintenance Representative 111
    Participation of Key Production Associates 112
    Participation of the Production Engineering Manager and Staff ...113
    Participation of Salaried Employees Detached from Production ...114
    Participation of Local Union Officials 114
    Preparatory, Wrap-Up, and Follow-Up Aspects of a High-
    Impact Kaizen Event 118
    Basic Structure and Steps Involved in Conducting a High-
    Impact Kaizen Event 119
Getting the Most Out of Training and Implementation Kaizen 121
    Modifying the Rules for the Purchase of New Equipment 124
    Planned Frequency of Training and Implementation Kaizen
    Events 125
Driving the Use of Problem Resolution Kaizen 126
    Applying the Science of 5-W 127
How to Conduct a Problem Resolution Event 128
Essential Tools Utilized in a Problem Resolution Event 131
Keeping the Principles of Uninterrupted Flow and Workplace
Organization in Mind 132
Understanding the Role and Scope of Sustaining Kaizen 133
Implementing a WRAP Initiative 135
    When to Start a WRAP Initiative 135
    Planning Phase Considerations 135

Typical Hurdles to Clear...138
Summary Overview of the Process..139
   Communicating and Tracking Results...............................139
Training First-Line Supervisors...142
Key Summary Points...142

5 Other Key Facets of Getting the Most Out of Kaizen...............145
Advancing the Role of Owner-Operator to "Lean Equipment
Specialist"...145
   Where LE Specialists Should Be Considered 146
   Pay Structure for a Lean Equipment Specialist Classification........147
   Percentage of Workforce Holding the Classification.....................148
   Lean Equipment Apprentice Training.................................149
Conducting an Annual Structured Lean Audit............................152
   Sharing the Results of the Audit with the Workforce 153
Building in Essential Visual Controls 154
Constructing a Master Kaizen Plan 156
Vendor Certification 157
Ten Commandments of a Fully Supportive Maintenance Function......159
Briefly Addressing the Cost and Payback of Lean Again 160
Ten Most Important Factors to Keep in Mind............................161
Phase One: Setting the Stage..161
   FACTOR ONE: Learning to Trust the Process...............................161
      For the Plant Manager 163
      For the Lean Coordinator...164
      For the Production Manager 164
      For the Shop Floor Supervisor 164
      For the Production Engineer....................................165
   FACTOR TWO: Assigning Appropriate Talent..............................165
      A Bit More about the Production Manager Position.............166
   FACTOR THREE: Doing the Planning Required to Put ALIP in
   Place...169
   FACTOR FOUR: Applying Strong and Effective Communications 169
   FACTOR FIVE: Demonstrating the Process....................................170
Phase Two: Completing the Mission 170
   FACTOR SIX: Training the Production Workforce 170
   FACTOR SEVEN: Driving Good Lean Practices into Office
   Processes 170

FACTOR EIGHT: Advancing Improvements at an Individual Job Level...171
FACTOR NINE: Applying Lean-Oriented Vendor Certification Standards 171
FACTOR TEN: Focusing on Continuous Improvement..................171
A Final Word 172
Key Summary Points..172

Appendix A: Recommended Reading **175**

Appendix B: How to Obtain Direct Assistance with the Process 177

Glossary: Definitions of Frequently Used Terms........................... 179

Index **183**

List of Figures

Figure 1.1   Conversion to Lean: Toyota original versus ideal..................... 6

Figure 1.2   When, where, and to what extent?... 8

Figure 1.3   ALIP: Advanced Lean Implementation Process....................... 9

Exhibit 1.1   Key differences in approach...11

Exhibit 1.2   Kaizen evaluation form. 18

Figure 1.4   Progressive Kaizen. 29

Figure 2.1   Pie chart: Distribution of Lean accomplishments....................37

Figure 2.2   The F Alliance..45

Figure 2.3   Example of appropriately weighted objectives. 50

Figure 2.4   Lean implementation step chart. 57

Figure 3.1   Conventional versus world-class manufacturing. 75

Figure 3.2   Production engineering—Conveyance of talent and ability. 79

Figure 3.3   Versions of a process.. 89

Figure 4.1   Progressive Kaizen component tool box. 97

Figure 4.2   Factory conversion: "The Swiss cheese phenomenon."........... 99

Figure 4.3   Inclusion of separate Kaizen labor pool. 117

Figure 4.4   High-Impact Kaizen event..120

Figure 4.5   TI event window diagram. 122

Exhibit 4.1   Example: Lean Manufacturing equipment checklist..............125

Exhibit 4.2   The inherent benefits of "Why?"...128

Figure 4.6 PRK event window diagram. 130

Figure 4.7 Example of time compression (Uninterrupted Flow). 132

Figure 4.8 The components of Sustaining Kaizen. 134

Figure 4.9 WRAP introduction timeframe. 140

Figure 4.10 Twofold objectives of WRAP. 141

Figure 5.1 Owner-operator training window diagram. 149

Exhibit 5.1 Electronic final assembly Andon board. 154

Exhibit 5.2 Example of simple visual controls. 155

Figure 5.2 First phase of 10-step roadmap. 162

Figure 5.3 Second phase of 10-step roadmap. 162

List of Tables

Chart 2.1   Step Chart Phase 1: Lay the Groundwork to Change the Factory's Production Technique (Timeframe: 8 to 10 months) 58

Chart 2.2   Step Chart Phase 2: Change Flow to Best Accommodate Lean and Pull Production (Timeframe: 6 to 8 months)..58

Chart 2.3   Step Chart Phase 3: Fully Implement Lean Manufacturing Throughout (Timeframe: 9 to 12 months) 59

Preface

Although there is little new that could be added in establishing industry awareness regarding the benefits of Lean Manufacturing and the tools of the Toyota Production System, a need exists to understand how to put it all together and fully implement the process in the most effective and least disruptive manner.

The United States is currently struggling through the worst economic downturn since the Great Depression. General Motors, once the largest and most influential car manufacturer in the world, filed for bankruptcy in 2009, and along with Chrysler has been struggling through a significant downsizing and restructuring. Ford has avoided doing the same; however, the automaker reportedly lost $10 million in 2008.

On average, America's manufacturing sector hasn't fared a great deal better. In fact, recession aside, U.S. manufacturing has been in a steady downward spiral since the advent of NAFTA, which made it advantageous for companies to relocate manufacturing to emerging economies, where the cost of labor and associated benefits are substantially lower. However, putting aside the things manufacturing essentially has no real control over, the question becomes what can be done to help firms achieve and maintain a strong competitive position for the future.

There are those who say the United States needs to look to the future for any stronghold it hopes to achieve in something other than manufacturing. Such talk is built on the feeling that we've essentially been beaten at our own game. Although that perception is correct to some degree, what isn't correct (and far from acceptable) is giving up without a truly concerted effort. America vitally needs a strong manufacturing base and, as pointed out, there's indeed a way to substantially strengthen our competitive ability. But in order to properly set the stage, a little history is warranted.

In the early 1980s Toyota became universally recognized as having founded a new system of production that greatly reduced inventory, lowered operating costs, and substantially improved product quality and delivery among other significant advantages. As a result the world of competitive manufacturing changed, never to be the same again. Toyota's system, which actually came about over a 40-year period of trial and error, was a complete reversal from the more traditional way of producing products and meeting customer demand. Although America was initially slow to the calling, it has since been fighting back with a process that's come to be labeled *Lean Manufacturing*.

Outside of the common tools used in applying the process (SMED, Poka-Yoke, Kanban, pull production, etc.)* the application of Lean means many different things to the managers, shop floor supervisors, production associates, and others who make up America's manufacturing workforce. In addition, the forces that serve to drive performance and the mind-set of employees haven't been fully addressed in support of the effort.

There will always be distinct differences in the manner firms approach basic operating procedures, but there is a level of commonality associated with the fundamentals involved. Every manufacturer holds the objective of producing and delivering goods at an acceptable price, while remaining responsive to ever-changing competitive pressures. The issue becomes how to introduce needed change in a manner that not only allows effective progress to be made, but includes a mechanism that strongly encourages the direct participation of the entire workforce.

As someone who has held the position of plant manager, I completely understand the difficulties associated with striving to incorporate change as complex and all-encompassing as a complete revision to a factory's system of production. But that is precisely the task U.S. industry faces if it hopes to achieve and maintain a viable competitive position for the future. If one accepts the odds are low of being significantly better than the competition in acquired talent and the procurement of needed materials and components, the principal area of focus left is a company's system of production.

A firm can always strive to meet competitive pressures with better equipment and product design, but such an advantage typically doesn't last. However, work at incorporating a superior system of production and history has shown the competition will usually be slow in responding, if they ever indeed fully and effectively respond.

One of the best examples is how long it has taken the United States to collectively buy into the principles and concepts of the Toyota Production System (TPS) and make it more than an industry watchword. Even as Lean

---

* See Glossary for definitions.

Manufacturing grew in awareness, the United States was slow to respond and in fact is still striving to catch up. But catching up with a superior system of production that places a persistent focus on consistent improvement, as an integral part of its operating philosophy, will almost always leave the competition a step behind. The reason why is because correctly incorporating all the principles and concepts involved in a complete revision to an operation's system of production goes against what most managers, supervisors, and others have been formally educated and oriented to do.

The stigma of formal education pertains not only to those who manage and direct manufacturing operations around the United States, but also our nation's source of higher learning. Dr. Mark A. Curtis, Vice President for Instruction at Alpena Community College and past professor at Purdue University, very astutely noted in an initial review of this work:

> The reason production engineers have not been at the forefront of the Lean transformation is it runs counter to their training in college and the professors' training which predates the whole Lean deal by 20–30 years. As a professor myself I was trained in reduction of direct labor, transfer machines, bigger and faster machines, economic order quantities that considered the cost of getting product versus the cost of having product. All this goes out the window with Lean, but it is hard for a professor to turn his back on years of college training and real-world experience that did not include Lean thinking. (personal communication)

Dr. Curtis went further to say,

> Some plant managers have risen to the level of plant manager because of their success and competence with the old batch and queue system. In short they are good at the old way and must be convinced of the benefits of Lean and then learn about it. Situational leadership would say not all leaders (plant managers) will be good in all situations. (personal communication)

America began to lose the advantage of being a manufacturing entity unto itself in the early 1980s, bound by a system of production built on the fundamentals of high-volume batch production. As time went by, the system became influenced by sophisticated storage and transfer systems, by highly paced production lines, and by throughput standards that were aimed more at theoretically covering every problem that might arise, rather than focusing on establishing the root cause of recurring production problems and putting them to rest.

The by-product perpetuated extremely large levels of inventory, established to keep the wheels of production churning. Simple but important things, such as long setup and changeover times, were completely ignored and associated wastes such as scrap, rework, and obsolescence were accepted as the cost of doing business. Most large manufacturing firms ended up with literally millions of dollars tied up in huge inventory storage and smaller operations had a similar abundance, in terms of a ratio of inventory to sales. Inventory became viewed as an operational asset, outweighing in importance the cost associated with carrying, managing, and handling the overabundance, even when sales did not reflect the need.

But inventory wasn't the only issue that came to influence a decline in America's manufacturing competitiveness. The ever-escalating cost of labor was another, along with the fact that standard performance measurements (such as machine utilization, direct labor efficiency, and others) motivated output over any other sense of direction. The basic motto essentially boiled down to: keep operators and equipment busy building parts, regardless if there's an immediate need or not. Anything more than required to meet customer demand can always be stored for future use.

This approach was workable as long as the competition boiled down to those using the same mode of operation. But as the world of business and subsequent competition began to rapidly expand beyond America's shores, a new approach to manufacturing emerged with a completely new set of operating principles. However, manufacturing in the United States initially turned a deaf ear to the calling, even when it became apparent it was fighting a losing battle; and the facts are we still haven't collectively turned away from the practices of the old and fully embraced a proven and more effective mode of operation.

There are two overriding reasons why this has occurred. The first concerns the lack of a mechanism that promotes the type of change required and serves to keep things moving in the right direction. The second, as previously mentioned, has to do with the difficulty of making massive change that's in direct conflict with an individual's formal education, training, and work experience.

The job to be done should not be underestimated. It is massive, in terms of not only physically changing the factory floor but also the mindset of plant leadership and the workforce as a whole. Chipping away at the insertion of Lean Manufacturing, much like Toyota was forced to do over decades of trial and error in developing TPS, simply will not suffice. An interesting thought to ponder is whether Toyota, given what they know today, would follow the same path they initially took in making the Toyota Production System a full reality. It's reasonably safe to say they would go about it entirely differently, especially considering the full creation of TPS was well over 30 years in the making.

The principal issue American industry faces isn't how to use the tools and techniques developed in the Toyota Production System. Most companies striving to apply Lean have gotten quite good at doing so. Again, the challenge centers on how to put it all together and aggressively go about making a full and complete transition, while responding to performance standards that in many ways pose a direct conflict with that mission. This has proven to be a delicate balancing act for most U.S.-based companies; but there is indeed a way to deal with the dilemma, which will not only aid tremendously in advancing the insertion of Lean but has the potential of making Kaizen a valid competitive weapon. However, entire thinking about Kaizen, its role, and its use has to change.

An excellent definition of Kaizen can be found in Wikipedia (the Internet-based encyclopedia) that states:

> Kaizen is a philosophy focusing on continuous improvement in manufacturing activities, business activities in general and even life in general, depending on interpretation and usage. In Toyota Kaizen is a daily activity, the purpose of which goes beyond simple productivity improvement. It is also a process that, when done correctly, humanizes the workplace, eliminates overly hard work ("muri") and teaches people how to perform experiments on their work using the scientific method and how to learn to spot and eliminate wastes in business processes. While Kaizen usually delivers small improvements, the culture of small improvements and standardization yields large results in the form of compound productivity improvements.

It would be beneficial for most manufacturing firms to carefully study that definition and strive to take it to heart because there are two aspects that typically are not being applied to the fullest. The first has to do with making Kaizen a daily activity. The second involves teaching people how to perform experiments on their work and how to spot and eliminate wastes in business processes. I would point out the words used were "daily," "people," and "business processes," which refer to more than just shop floor operations and hourly production workers. The business of waste reduction can and should involve every area of the business, including office functions, which haven't been completely ignored but ideally need a great deal more attention.

Kaizen was added to Toyota's war chest after a great deal of fundamental work had been completed on the shop floor. Taiichi Ohno, the recognized father of the Toyota Production System, said as much when quizzed by Norman Bodek, a forerunner in the research of TPS and its growth in Japanese industry. Bodek proceeded to ask Ohno where the company stood "after reducing all work-in-process inventory, lowering the water level to expose the rocks and enabling them to chip away at the problems." Ohno's response was: "All we are doing is looking at the time line from the moment the customer gives the order to the point we collect the cash. And we are reducing that time line by removing the non-value-added wastes."[1]

What Ohno was essentially saying was that Toyota was beginning to focus strongly on continuous improvement, in order to complement the fundamentals they had worked to put in place over the course of many years. Our opportunity lies in using an energized Kaizen process as the "accelerator" for a much more thorough application of Lean.

A vital task we face in turning the attention level up dramatically rests with somehow changing the thinking of the average manufacturing manager, who from both an educational and experience standpoint has been strongly oriented toward conventional manufacturing practices. Changing such a mind-set isn't easy, especially in light of the fact that many leaders who fit the category have been highly successful utilizing the techniques they came armed with out of college and which were further enforced with their personal work experience. Asking them to consider changing the techniques they've previously been successful with for years doesn't always sit well. As Dr. Curtis pointed out, something has to serve to convince them thoroughly of both the need and the way.

I was no different. I grew up in a batch manufacturing environment, assumed the role of plant manager under that scenario, and had to literally be pushed by a strong mentor to find out more about the benefits of Lean. I rather reluctantly attended a training course that I honestly wasn't all that enthused about. True to my commitment, however, I arranged to attend a day-and-a-half seminar conducted by Richard J. Schonberger, an early author and lecturer on the subject of world-class manufacturing. I asked a number of my staff, along with the leadership of the local union, to accompany me to the seminar and we collectively came away with an entirely different view about manufacturing.

As time went by, I became a strong convert and energetically went about applying everything I could learn about the process, which resulted in a factory that became a showcase operation for United Technologies. I went on to spend four years traveling and lending assistance to manufacturing operations, conducting numerous (two-week) "high-impact" Kaizen events in UTC facilities around the world.

If I was capable of changing my mind, anyone should be able to. I went to work in manufacturing at the age of 24, making my way through the industrial engineering ranks, before taking the job of plant manager at the age of 48. I had over 20 years of conventional manufacturing practices literally pounded into me before coming to recognize the need for change. I can therefore say to any manager who may have doubts about Lean: open your mind to a trusted and proven process and do everything possible to make it a reality. You will come to applaud the day you made the decision.

Using Kaizen to the fullest extent can aid tremendously by making it the chief mechanism for the type of change required. But we have to learn to perform Kaizen in a much more effective manner than has typically been the case, which starts by understanding what the process is actually capable of accomplishing.

Most Lean initiatives lack definition, with respect to how they fit in the overall scheme of achieving a clear end objective. That is why a reasonable amount of formal planning is required before extending efforts that basically become a haphazard approach to the task. Although it may be nice and somewhat comforting to put various facets of Lean Manufacturing in place in a factory (which can be pointed to as being on the right track), it achieves little without a thoughtfully structured plan to make a whole and complete transition in the manner production work is performed. Therefore, the planning side of the equation becomes essential. But a viable and constructive plan of action cannot be duly accomplished without first committing to an engine that serves to drive the entire process.

A good Lean Manufacturing effort requires actions that serve to correct established paradigms, as well as production processes on the shop floor. This cannot be done in an adequate manner without appropriate training and communications, along with a well-identified accelerator for the process. This book points out a means to transition the concept of Kaizen from a less than frequently used tool—for the most part aimed at making small change under the guise of incremental improvement—to an all-encompassing process aimed at driving strong lasting change through an entire factory.

Although one can go about placing a turtle on a downhill slope in order to temporarily increase its speed, the fact is that it's still a turtle and in a competitive race with almost any other four-legged creature it's absolutely no match. To some degree a comparison can be made to the manner in which manufacturing firms have basically approached Lean. The focus has been on making small incremental improvements to a slow and cumbersome waste producer, rather than working on getting that waste producer completely out of the picture and replacing it with something with much greater speed and endurance.

Getting out of the rut of making Lean all things in theory and very little in overall competitive practice (i.e., fully eliminating an outdated system of production) is essentially where the opportunity rests. The use of the tools of TPS and the philosophy that serves to drive them have come to be known throughout the United States as Lean Manufacturing. It's a term sometimes inadvertently taken to mean being lean on employment. But although Lean is structured to address the best utilization of employees and frequently calls for fewer operators than typically used to do the work involved, employee reductions are not the chief objective. In fact, if applied in an effective manner, the long-term result elevates the potential for added employment and much greater overall job security.

The primary objective of any Lean effort should be to make manufacturing more responsive to the customer, while at the same time removing inherent wastes that serve to increase costs and reduce overall flexibility. In an effort to make the task more understandable I outlined the fundamentals involved in *Fast Track to Waste-Free Manufacturing,* in terms of four guiding principles that the average American worker could easily relate to and rally around:

Guiding Principles	Toyota Production System Tools
Workplace Organization	5-S, Visual Controls, Std. Work, U-Cell
Uninterrupted Flow	Pull Production, Point-of-Use Mfg., Kanban
Insignificant Changeover	SMED, 5-S, Visual Controls, Std. Work
Error-Free Processing	Poka-Yoke, TPM

But to avoid any confusion as to the connection between this and my previous works on the subject of Lean, the following provides a quick summary.

Fast Track to Waste-Free Manufacturing was aimed at explaining the benefits of Lean and boiling down the fundamentals to a set of guiding principles that made the task more understandable for management, shop floor supervisors, and hourly production employees. *Leading the Lean Initiative* was designed to address the many issues plant managers typically face in making a transition to Lean, along with the role they should play in the overall effort. *Lean Manufacturing; Implementation Strategies That Work* focused on a formal plan of action aimed at adequately engineering a factory's key production equipment, in order to more effectively support a Lean initiative.

The question could arise as to why further work was needed and indeed warranted. Again, the answer lies in the fact that although the process of inserting the principles and concepts of Lean has begun in earnest, in numerous manufacturing facilities across America, there isn't a universally accepted mechanism that serves as an accelerator for the process. Kaizen can indeed be that accelerator if it is fully understood and used in the correct manner, thus the purpose of this book.

Holding what might be viewed as an excellent Kaizen event is actually next to useless if the gains are not carried forward as part of an ultimate plan to revamp the entire factory. Unfortunately, completely revamping a factory isn't the common mission of most Lean/Kaizen efforts. Most are chip-away-oriented and aimed at short-term cost reduction. Although it could be argued that the ultimate result of such efforts have gone a long way at fully incorporating Lean, a problem becomes apparent in the definition of the task:

- Fully incorporating Lean means entirely eliminating any form of batch production. In order to accomplish this and overcome the influence of "process monuments" (e.g., a large paint line or coating process that can be extremely expensive to relocate or replace), a well-defined Kanban process has been established and rigidly enforced. Any factory that has not accomplished the full insertion of "pull" throughout its entire production process simply has not fully incorporated Lean.
- Fully incorporating Lean means that each and every piece of key production equipment has had a well-qualified application of SMED, Poka-Yoke, and TPM thoroughly applied and that nothing is done that doesn't fall within the parameters of a set of guiding principles. When something fails to meet those guiding principles, great care is taken before pursuing it to completion.

- Fully incorporating Lean means making change and then following up with continuous improvement activities, conducted on both a formal and informal basis, through structured group activity and participation at an individual job level.

The old tired and worn cliché that Lean is a never-ending process has to be taken out of the equation. Certain aspects of continuous improvement fit that particular description; however, putting a solid foundation for Lean in place doesn't. But in order to build a proper foundation, it's important to remember there's a common barrier that has to be overcome. That barrier centers on the fact that even under the best of circumstances, individual goals and typical performance measurements are commonly constructed to support the old (existing) system of production, making it difficult to drive the kind of change needed across the entire factory. Thus, the accelerator of change has to be something that's capable of overcoming this particular barrier in an effective manner.

This book serves to address how to make Kaizen a formidable competitive weapon. Describing the end result in terms of being *formidable* could be viewed as a stretch, inasmuch as the word itself is defined in dictionary.com "of great strength; powerful and intimidating, arousing feelings of awe or admiration." But if one considers *formidable* to be a valid description of the Toyota Production System (an operating concept universally recognized for its manufacturing excellence), the outcome that can be achieved with the proper application of strategy and tactics is actually far from a stretch. Understanding the potential scope of Kaizen and how to most effectively apply it offer a means of achieving a competitive likeness to TPS that is second to none and that, by anyone's definition, would truly be a formidable competitive weapon.

Endnote

1. Ohno, Taiichi. 1988. Publisher's Foreword in *Toyota Production System— Beyond Large Scale Production*, New York: Productivity Press.

Acknowledgments

A special word of thanks to Michael Sinocchi, Senior Acquisitions Editor for Productivity Press. Michael was instrumental in bringing my first work on Lean Manufacturing to publication (*Fast Track to Waste-Free Manufacturing*); followed two years later by a second book (*Leading the Lean Initiative*). I brought the thought for this work to Michael at a time when the overall economy was teetering, affecting the publishing world as well as almost every other facet of business. It was also a time when a massive amount of material abounded on Lean and Kaizen. But Michael saw something uniquely important about the subject matter and urged me to bring the project to fruition. For his continuing support and encouragement he has my sincere appreciation.

I would further like to recognize Roger Lewandowski, CEO of World Competition Consultants and a longtime friend and mentor. During his career with Carrier Air Conditioning, he served as Vice President of Manufacturing for Carrier-North America and was instrumental in providing me with my first opportunity in a plant management position. He went on to serve as a continuing advisor over the course of my career and is someone I hold in the highest regard. Without his strong and lasting influence I simply would not have been exposed to the background and experience I was fortunate to gain.

Last but not least, I would like to express my gratitude to Barry O'Nell, Edward Cannon, and Gary Roscoe. These gentlemen worked with me on a very special assignment conducted between 1993 and 1997, aimed at deploying *Flexible Manufacturing* for United Technologies Corporation. Over a four-year period we traveled much of the world, conducting two-week high-impact Kaizen events at facilities in the United States, South America, Europe, and the Far East. It became the catalyst for a wealth of knowledge and hands-on experience in Lean Manufacturing to which we would not have otherwise been exposed. I'll always be exceptionally grateful for the unwavering commitment and support they extended in making the effort a truly professionally rewarding experience.

Introduction

The inspiration for this work came shortly after finishing *Lean Manufacturing; Implementation Strategies that Work*. I was having a conversation with a friend and colleague, who like myself had gone into consulting work after a long career in manufacturing. Although our fields of endeavor were different, his in accounting and mine initially in industrial engineering, we shared many common views about manufacturing.

As the conversation went on he mentioned that one of his principal frustrations with Lean rested in the fact there seemed to be a great deal of trial and experimentation involved, with little actual return on overall investment made. I told him that I essentially agreed, but the culprit wasn't Lean itself. I went on to say what was missing was a lack of a solid approach to implementation and the surest way of getting there was elevating the use of Kaizen.

He confessed he didn't fully understand what I was driving at and in response I asked him to accompany me to my office, where I took the time to sketch an outline of a concept I refer to as "ALIP" (Advanced Lean Implementation Process), which utilizes "Progressive Kaizen" as one of the three major components. When I finished he studied the whiteboard for a moment before remarking, "You know something, John? You need to write a book about that."

Although I appreciated the comment, I thought little more about it until three weeks later when he called and asked how the book was coming. I had to admit to him that after just finishing a book on Lean, I hadn't given a lot of consideration to applying the necessary time and dedication required to bring another work to fruition. His response was: "You know best about that, but I think it's something manufacturing firms could really use."

His persistence spurred me to sit down and construct an outline for a manuscript and as it started to take form it became increasingly apparent that it was not only a highly worthwhile topic to pursue, but one that held the potential of greatly enhancing the progress of Lean across most U.S. industries. The reason I go as far as saying that is because of the unique subject matter this book serves to address.

There's indeed a bountiful amount of material available that speaks to Lean and Kaizen; however, this work is a first in breaking down the process into four distinct categories of application, clearly pointing out that all Kaizen activity is far from being one and the same. It comes with almost 30 years of experience in a wide diversification of manufacturing, and from someone who lived through the challenge of applying Lean as a plant manager in two very large and complex factories that were laden with waste and inefficiency. I therefore feel the outlined concept can be extremely helpful to any manufacturing operation wishing to advance a Lean initiative, and especially for those who hold an interest in getting the utmost out of Kaizen.

Who Can Benefit from This Book and Why

Although there is little that can be added regarding the various tools and techniques of the Toyota Production System that hasn't been addressed in a massive amount of available material on the market, this book offers a unique process aimed at building uniformity in the way companies can more successfully go about fully implementing Lean. The targeted emphasis is on an advanced application of Kaizen that serves to promote an increased level of awareness and workforce participation. In keeping with this, the functional role and responsibility of key positions are thoroughly addressed and the various types of Kaizen are identified so it is easy to understand their specific use, along with when and how these can be applied in the most effective manner.

For the College Professor and Engineering Student

The content provides a teaching and study aid that can be used by college professors and engineering students in the realm of Lean Manufacturing and the supportive role graduates should come prepared to play when they enter industry. Two of the more important topics addressed center on the responsibility production engineers should hold in upgrading equipment to support a Lean initiative using an advanced application of Poka-Yoke, SMED, and TPM,* along with the role they hold in providing input and assistance in Lean applications to production supervisors, shop floor employees, and others as conditions warrant.

For the Plant Manager and Executive Company Leadership

The content can serve as a roadmap for a process that can lead to making a factory, a company, and an overall enterprise much more competitive to the challenges of the future. Whether adopted in full or in part, the use of the techniques, concepts, and strategies involved can be highly beneficial in advancing the implementation of world-competitive manufacturing practices and can do nothing but improve a manufacturing operation's overall strength and ability. The content involves a structured guideline for getting the absolute best out of Kaizen and taking a Lean initiative to its ultimate level of achievement.

For the New and Experienced Lean Coordinator

This book serves as reinforcement for the key role the Lean coordinator holds in training and leading change that serves to make and keep a

---

* See Glossary for definitions of terms.

manufacturing firm world competitive. The content provided can help the Lean coordinator in structuring a master plan that elevates the overall use and effectiveness of Kaizen and further promotes a much more effective approach to the task of fully implementing Lean in keeping with the four guiding principles of workplace organization, uninterrupted flow, error-free processing, and insignificant changeover.

For the Company or Operation Just Entering or Considering a Lean Initiative

The content explains and reinforces the importance of Lean Manufacturing and how to construct a change process using Kaizen as the chief mechanism for implementation. It further speaks to the various types of Kaizen activity, the associated role of key players, needed organizational change, and how to avoid common pitfalls. It can be viewed as a roadmap to more successfully approach the task of implementation and, as the subject requires, with an abundance of visual aids and applicable reference material.

For the Production Manager and Shop Floor Supervisor

The content explains the importance of fully implementing Lean Manufacturing practices in a factory and the specific role the production manager and shop floor supervisor should play in thoroughly supporting the process and making it a lasting reality. Included are the types of actions and individual objectives that should be assumed.

For the Employee Desiring to Increase His or Her Overall Knowledge and Expertise

The content speaks to a growing need in manufacturing and the opportunity for employees (both hourly and salaried) to increase their overall knowledge in a field that will only continue to expand in importance as time goes by. Effectively learning how to use the principles and concepts involved doesn't require a college degree and is commonsense oriented. What is ideally required, however, is striving to work for a company that utilizes Lean Manufacturing and provides both training and direct involvement in the process. Once the skills are learned and practiced, the acquired expertise will substantially increase an employee's value in today's swiftly changing manufacturing environment.

Key Supporting Material

There are two books frequently referred to throughout the content of this book. One is my first work on Lean, *Fast Track to Waste-Free Manufacturing*, and the other is a more recent one entitled, *Lean Manufacturing; Implementation Strategies That Work*. Although what is outlined herein is essentially self-supportive in providing a viable approach to Lean and an expanded use of Kaizen, in order to obtain the most comprehensive understanding of two key topics addressed, these two books are highly recommended reading. The first key topic involves the expanded role of the production engineering function, covered in depth in *Lean Manufacturing: Implementation Strategies That Work* (Davis, 2009). The second key topic addresses four guiding principles which serve as the foundation for Lean implementation, covered in expanded detail in *Fast Track to Waste-Free Manufacturing* (Davis, 1999).

Chapter 1

Examining the Basics of an Effective Kaizen Process

When I've gone about asking various manufacturing firms the specific role Kaizen plays in their operation, I've gotten a variety of answers. Unfortunately, the one I've failed to hear and that most appropriately fits the task is: "Kaizen is the chief mechanism used to fully incorporate Lean and provide employees a means of making improvements to their individual jobs."

In all fairness, however, most manufacturers haven't focused on establishing a formal strategy and a well-defined path for Kaizen. Accomplishing this starts by understanding that Kaizen can be much more than an instrument for small "continuous improvements," the thing it's chiefly touted to accomplish. In reality there are four distinct types (or categories) of Kaizen, each with its own purpose and priorities. A company that learns to identify with these and use them to their ultimate can make gigantic steps in the right direction.

There are few who would disagree that the minimum goal we should be striving to accomplish is to bring our manufacturing expertise up to par, as quickly as possible, with Toyota and others who have adopted a superior system of production. However, the level of change simply hasn't gone far enough in most cases. Although the United States will probably never overcome the disparity that exists in labor costs and benefits, it has the chance of much better managing and controlling the other major facets of the business: more specifically plant overhead, indirect labor, both work-in-progress and finished goods inventory, along with reducing manufacturing lead time and increasing overall responsiveness to customer needs. But getting that point across to those who direct and manage manufacturing operations in the United States can sometimes be difficult.

RELATED EXPERIENCE: I was in the process of striving to explain this to a group of mid-level manufacturing managers when one of them interrupted and said, "Well if we can do it, so can the competition. That still leaves us behind the eight-ball." I could see an immediate flash of agreement sweep across the faces of the others in attendance and it became apparent I had to swiftly deflate a growing defeatist attitude or I was otherwise going to lose the audience, both in mind and spirit. But changing the mind-set of a group of highly seasoned managers, who have seen enough to be convinced that the chances of a silver bullet are slim, is difficult at best. I clearly understood I had to choose my response carefully and admittedly had to think a moment before replying. But I did so with two questions, a short story, and a point.

The first question: Did the audience believe Toyota was a formidable competitor in the automotive industry?

The response: A collective "yes."

The second question: Were they aware Toyota almost went bankrupt and stood the chance of going out of business and closing the doors to its factory forever?

The response: "No" to "Not entirely."

Capsule of the story: In the mid-1940s Toyota was just coming out of World War II and struggling to enter the automobile industry. Things didn't go well, to say the least. Toyota found itself on the verge of bankruptcy and facing the looming possibility of closing its doors forever. Deming and Duran came on the scene and Toyota became a forerunner in eagerly adopting statistical process and quality control procedures that served tremendously to keep the business going. Later, under the guiding influence of Shigeo Shingo and Taiichi Ohno, the birth of what today is known as the Toyota Production System took place and the rest is history.

The point: Toyota could have easily said, "If we can do it, so can the competition," and gone about throwing their hands up in defeat. But they didn't and today they're viewed among the titans of the automotive industry. Why? Because Toyota was willing to put aside conventional thinking and pursue an approach that drove typical waste and inefficiency out of manufacturing. I went on to stress to the group that it shouldn't be assumed because the same opportunities exist for the competition that they will take a genuine interest in eagerly applying the time, energy, and effort to make it happen. American industry certainly didn't when it had the opportunity to follow Toyota's lead decades ago, and there are those today who still aren't fully convinced of the urgency.

In the end the group developed an acceptance of the opportunity and turned into a highly energetic team that went on to lead some outstanding accomplishments for the factory over the course of the following three years. With only minor reductions in direct labor, the plant focused on overhead and managed to reduce what is referred to in the financial branch of the business as "burden" from 330% to 260%, a 70 percentage point improvement. Considering inflation, it was indeed a phenomenal accomplishment that went on to favorably affect profitability in an industry where pennies counted when it came to selling price.

There are two reasons for relating that experience. The first has to do with pointing out the problem of striving to sell the urgency for change to those who hold the responsibility of managing the day-to-day activities of manufacturing. There will always be doubters, especially if the product happens to be a concept, because it is extremely easy to dismiss the thoughts and opinions of others if the buyer doesn't have a product they can see, feel, and touch.

The second reason is to provide an example of a wide variety of personal experiences that will be shared throughout the content of this work. These are highlighted in shaded boxes should the reader desire to flip through and quickly refer to them. Collectively, they involve many issues that frequently arise when striving to implement Lean, taken from a wide range of both career and consulting experiences. But if there is one thing the reader should keep in mind and carry with them long after setting this book aside it would be this:

Kaizen isn't limited to the single purpose of making small continuous improvements. Used in the correct manner, it can serve as the

chief mechanism in fully inserting Lean Manufacturing throughout an entire business enterprise.

Learning how to get the proper kind of focus and how to construct activities that serve to get the best out of Kaizen is what the content of this work is designed to cover. Some of the more prominent issues that are addressed center on:

1. Defining the four basic types of Kaizen and their intended goals and objectives.
2. Outlining the characteristics of a select group of leadership roles that allow a firm to get the best out of Kaizen.
3. Providing the specific dos and don'ts associated with making Kaizen all it can be.
4. Defining the critical role the maintenance function plays in Kaizen and what to do to strengthen its ability to respond in a more effective manner.
5. Establishing the merits of a WRAP (Waste Reduction Activity Process) aimed at effectively driving Kaizen down to the individual job level, in each and every branch of the business.
6. Pointing out how to go about avoiding the common pitfalls associated with Kaizen.
7. Outlining the key factors associated with developing a firm plan of action for Kaizen and seeing it through.
8. Addressing the insertion of effective measurements and reporting activities that serve to drive and steer the process and avoid stalls in implementation.

"System of production" is a term used frequently throughout the content of this work. In the simplest form of definition it refers to both the philosophical mind-set that drives the action and the specific techniques used to perform the work involved. Fully changing the system of production from batch, to something strongly representative of the full insertion of Lean, is the principal task involved. Much of what occurs in a typical manufacturing operation could be done with far less waste and without changing the existing system of production. However, the common measurements and individual objectives used to drive an operation often prescribe the absolute wrong practices; the most prominent error being overproduction and everything that precedes and follows this particular flaw. This in turn means the old system of production has to be torn down in its entirety and fully replaced.

To some degree we need to speak about change in terms that make the entire process something middle managers, floor supervisors, and production employees can more easily rally around. As an example, a more plausible definition for Kaizen would be *waste reduction activity*, because employees would typically be far more energized about participating in a process clearly aimed at reducing manufacturing wastes than they would in being asked to serve in a "Kaizen" event—which even today is still a bit of a mystery to most American workers. But more about this particular subject later.

Unfortunately, Kaizen is often something employees are not encouraged to actively utilize in their daily jobs. In many cases Kaizen turns out to be a special affair performed on a relatively infrequent basis. This has helped create the principal difference between what Kaizen means in Toyota versus what it means in U.S. industry. In the sense of the greatest difference, in the United States the objective is commonly aimed at picking the low-hanging fruit and achieving short-term cost reduction. Although there is nothing wrong with that in itself, we tend to lessen the overall power of Kaizen by not using it to fully change the existing system of production and then working to keep it that way.

If fully studied, history points to the fact that Toyota's principal use of Kaizen came after the basics of the Toyota Production System (TPS) were put in place. Our task is to put Kaizen on the front end, in order to more effectively make the kind of change required and then use it as Toyota has in supporting and maintaining a truly competitive system of production.

Figure 1.1 indicates such an approach, as opposed to the manner Toyota's system was developed over a period of decades. Under this scenario the very first task would be to fully engineer a plant's key production equipment to support a Lean process and then follow up by inserting the use of Kaizen as the chief agent for change. For those who already have a Lean Manufacturing initiative underway, doing this would essentially boil down to adjusting implementation strategy in order to see that the first step is fully carried out. Using this approach would serve to enhance any Lean Manufacturing initiative regardless of how long it has been in place and should not be viewed as changing the ultimate objective in any way. (See Davis's *Lean Manufacturing; Implementation Strategies That Work* for more detailed information on this important step.)

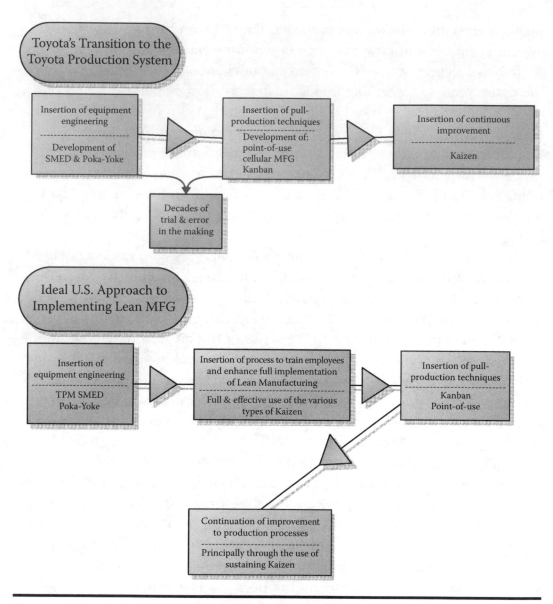

Figure 1.1 Conversion to Lean: Toyota original versus ideal.

For those who are into Lean and have a small or nonexistent production engineering resource, there could be some reluctance to hire the additional talent needed to fully carry out the equipment engineering noted. This, as pointed out, is shortsighted management. But regardless of that particular step, there should be absolutely no reason for not considering a strengthened and much more productive Kaizen process. At some point, however, a strong qualified application of Single Minute Exchange of Dies (SMED) and Poka-Yoke needs to be applied to key production equipment throughout the factory, and this is best achieved through the use of educated and fully qualified manufacturing and industrial engineers.

So what does utilizing Kaizen to its fullest mean in the application of day-to-day individual duties and responsibilities? It means taking the process to an increased level of activity, which involves well-trained operators on the shop floor who take it upon themselves to perform Kaizen as a positive influence on their work. It further requires production engineers who become teachers, advisors, and auditors of the process, along with shop floor managers and supervisors who have a knowledgeable respect for the process and strongly encourage their people to use it.

A portion of the content outlined is aimed at senior management and those who effectively lead a Kaizen effort. On the other hand, it's valuable reading material for any manufacturing manager or supervisor, along with those making up the general salaried and hourly workforce. Even for those who already have an aggressive application of Kaizen underway there are some compelling reasons to take the time to study carefully what is outlined, because:

1. It covers topics not typically addressed when speaking about Kaizen, but which are essential to the overall success of a well focused and well-run process.
2. It provides sound advice for those directly responsible for managing a factory and those assigned the duty of carrying Kaizen forward in a successful and meaningful manner.

Take a look at any Lean initiative and strive to identify the philosophical focus used to drive that initiative. In most cases it simply isn't there. What one typically finds is a rather confusing mix of tools and techniques, with no well-defined driver for the process other than when and to what extent management decides to pursue the tools involved (see Figure 1.2).

When most companies are asked to explain the philosophical focus that drives the effort, they will almost always reply, "Continuous improvement." The obvious question becomes: improvement to what—the existing system of production or something else; something higher and more rewarding or something aimed at striving to make an old and obsolete means of production somewhat more effective, overall? If there is any doubt about that, go to the leader of most any company or factory involved and ask him to define precisely what the work he is putting into a Lean initiative is aimed

Figure 1.2 When, where, and to what extent?

at accomplishing and what serves to drive the process. The response might surprise you.

Unfortunately, Kaizen is commonly viewed as just one form of ammunition in the arsenal of tools and techniques developed to implement Lean practices. Quite often the efforts extended hold the unwritten objective of building some, but far from all, of the advantages of Lean into a conventional manufacturing operation. This is due to many factors discussed in appropriate detail as we move along. But first and foremost on the list is failing to understand how *wrapping* the tools of Lean in an effective Kaizen process can move Lean implementation forward in a more productive and far less disruptive manner than normally the case.

Figure 1.3 provides a visual representation of how the outlined Kaizen process would ideally work. The tools and principles indicated will be discussed, but with a glance one can start to see what a truly results-oriented Kaizen effort can and should be. ALIP (Advanced Lean Implementation Process) is made up of three major components: Progressive Kaizen, Waste-Free Manufacturing, and WRAP. What becomes clearly evident is that everything is essentially driven by four guiding principles, initially outlined for waste-free manufacturing. This may appear to be a bit complex, but as shown, the entire process is common-sense oriented and easy to relate to once the elements involved have been thoroughly explained.

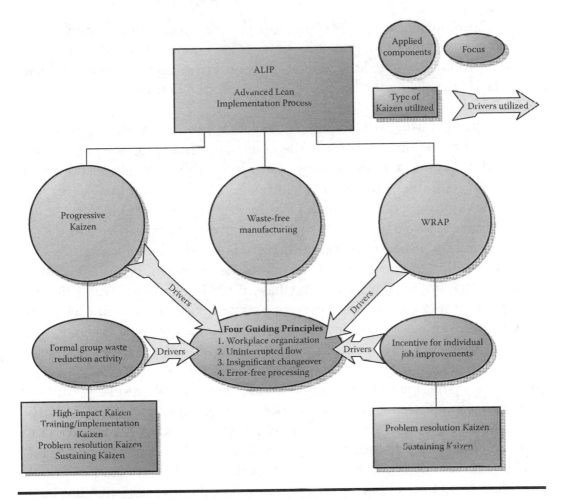

Figure 1.3   ALIP: Advanced Lean Implementation Process.

It would be helpful for the reader to highlight Figure 1.3 and keep it handy for quick reference purposes, because it essentially captures what this entire work encompasses. The overall concept of ALIP again starts with three applied components; Progressive Kaizen, Waste-Free Manufacturing, and WRAP. Specific information relating to WRAP is outlined in Chapter 2 under "Value of Inserting a WRAP Initiative" and again in Chapter 4 under "Implementing a WRAP Initiative." The categories of waste reduction noted for progressive Kaizen are addressed later in this chapter under "Overview of the Various Types of Kaizen" and are spoken to repeatedly throughout the content.

In a nutshell, the principles involved are Workplace Organization, which focuses on an efficiently constructed work area with adequate visual aids and controls; Uninterrupted Flow, which is aimed at substantially minimizing stoppage and storage points in flow; Insignificant Changeover, which places a focus on reducing setup and changeover to the point of it becoming insignificant in the decision-making process for taking on added business, revising production schedules, and so on; and Error-Free Processing, which is directed at applying the science of Poka-Yoke to eliminate the chance of recurring production errors or quality issues.

But putting a truly workable implementation process in place, through an advanced application of Kaizen as the accelerator, has to go a step further. This additional step requires taking a hard look at the role and responsibility of key players in the process and laying out a plan of action that serves to overcome typical roadblocks to success. As is addressed, creating an appropriate mind-set for change weighs no less in overall importance than the mechanism established to successfully drive and steer the effort.

In addressing the matter of perceptions, one of our shortcomings with Kaizen is that potential achievements have often been limited with thinking which says if it doesn't work to everyone's satisfaction, we can always revert back to the way we were doing things before. Kaizen is also frequently clouded with resistance aimed at avoiding disruption to the existing system of production, the very thing the process should be designed to eradicate. In many cases participants back away from clearly attainable results because of concerns about the support they will need to make a change in its entirety.

As an example, in order to get a "pull" system of production underway in a factory, the paint line supervisor may be asked to deliver finished parts to final assembly only in the quantities specified. In order to accomplish this, the paint line would be required to hold a portion of the inventory that has been "pushed" to them, due to no fault of their own, but because the plant as a whole is still operating under the guidelines of a batch system of production.

Although the paint line supervisor may not fight the change entirely, knowing Kaizen is being performed at the direction of senior management, he will sometimes construct compromises that serve to make the success of the change less than totally effective, such as obtaining an agreement with the group performing Kaizen to send only a select number of parts in the quantities specified and to continue business as usual on the majority of parts involved. This way the supervisor can say he hasn't resisted the change entirely. But in reality the effort ends up accomplishing next to nothing and the intended track toward the insertion of a true pull system for the factory essentially falls by the wayside.

If we're going to make effective change, we have to enlighten employees to the fact that the game isn't to accommodate the old system of production but rather to do everything possible to completely destroy it! If this means placing additional work and responsibility on various areas of the factory in order to drive the influence of Lean Manufacturing forward, it should be viewed as the price paid in getting there. However, getting everyone on board and having them energetically join in fully supporting the kind of change needed is one of the biggest hurdles to clear.

In order to properly set the stage, it's appropriate to briefly examine where the last three decades have driven the basic fundamentals of manufacturing. For the most part the fundamentals have shifted the focus from sheer volume to the ability to offer customers a greater variety of products, with better overall quality and deliverability. Toyota's system, which came about over many years of trial and error, was a complete reversal from the more traditional way of meeting customer demand. As a summary comparison of conventional batch manufacturing practices versus the Toyota production system, Exhibit 1.1 denotes the philosophical differences that exist and specifically where the four principles discussed earlier directly apply (see the underlined items).

The differences noted are not news to those who have a good understanding of Lean and the value it places on eliminating waste and inefficiency. But sometimes a reminder of the scope of the task involved helps

Conventional Manufacturing	Toyota Production System
Production areas work independently	Production areas work in unison
Production "pushed" onto next operation	Parts "pulled" from the source by user
Substantial setup and changeover	Insignificant setup and changeover
Large WIP inventory base	Vastly reduced inventory base
Work areas spread throughout factory	Work areas compact & flexible
Focus placed on overcoming problems	Focus placed on fully resolving problems
High scrap and rework levels	Essentially void of scrap and rework
Low solicitation of improvement ideas	Improvement ideas strongly encouraged
High operating costs	Extremely competitive operating costs
General flow has many stops & storage	Focus placed on uninterrupted flow
Poor workplace organization rules	Outstanding workplace organization
Frequent inherent processing errors	Focus placed on error-free processing

Exhibit 1.1   Key differences in approach.

put things in proper perspective. Unfortunately, I've seen numerous manufacturing operations where management was quick about declaring their accomplishments with Lean, but who were far from making the kind of overall change needed. Although they could always point to islands of improvement, it was extremely evident that many traits of a conventional manufacturer still lingered. Production areas for the most part still worked on an independent basis; production was still pushed through the factory, and work-in-process inventory levels still remained high, among other signs indicating no substantial change from the philosophical traits of old.

Admittedly, changing how things have been done for years on end isn't easy, especially considering that the old way of doing business was long viewed as the appropriate means of approaching the task. In fact, after World War II, Japan and others flocked to the United States to learn how to emulate America's approach to manufacturing. Joseph M. Juran and W. Edwards Deming saw some distinct flaws in America's system of production and went about developing quality procedures and statistical process controls to which America initially turned a deaf ear. Those same procedures were later aggressively adopted by the Japanese Union of Scientists and Engineers, and under the direction of Taiichi Ohno and Shigeo Shingo the entire process of manufacturing was taken to a new level, out of which grew the Toyota Production System.

But even with the growing knowledge that change was necessary, factories across the United States continued to be designed to accommodate a push-production mentality. Unfortunately the news media and others failed to help the situation, placing the entire blame on the lower cost of labor in emerging nations, rather than giving appropriate consideration to the basic system of production being utilized. It is indeed true that the principles of Lean Manufacturing have grown in acceptance, gaining steady momentum since the mid-1990s, however, the necessary change won't be accomplished to the fullest until our educational system and the people responsible for the nuts and bolts of manufacturing come together in a common mission.

For the plant manager who finds it hard to pull away from what has been highly successful in the past, I can personally testify that committing yourself to learning all you can about Lean will be extremely beneficial for both you and those who count on your leadership and direction. If you apply the same level of zeal and commitment to Lean that you've successfully applied to conventional manufacturing practices, you can rest confident you're on the right track. Don't stop if you are less than convinced at your first exposure to the process. Keep an open mind and stay committed to learning all you can. Go to more than one source and read the wealth of information that's available in today's market.

But the lack of a full commitment to change goes beyond the matter of swaying the thinking of senior management. As previously mentioned, an additional problem to overcome centers on common manufacturing measurements that drive a firm to do the absolutely wrong things. One such measurement is equipment utilization, which is theoretically aimed at reducing the part cost of a setup. Although this sounds efficient, the hidden expense comes in the form of inventory that has to be stored, counted, and cared for until the schedule calls for the parts to be used. This in turn leads to damage, scrap, and rework, and frequently unwarranted obsolescence. Firms have to come to realize what's important is ensuring that equipment is ready to run when it is needed, rather than focusing on how much time a machine is senselessly pounding out parts and unwarranted inventory. They also have to understand that it isn't only the added inventory that's costly; it's the bevy of other wastes that are compounded by the practice.

There are those who would strive to qualify everything that is currently going on by saying if it took Toyota decades to make the transition we can't be expected to accomplish the task in a relatively short period of time. That's the kind of thinking that has led us to where we are today. We can make fast and effective change by taking the best of TPS and putting the kind of planning and strategy in place around which the average production worker, manager, and supervisor can more energetically rally. For example, rather than extending a haphazard effort aimed at inserting each and every tool of TPS so it can be said they are being used, a much better approach is to use the tools only as needed in meeting the overall objective of implementing Lean, in each and every area of the factory.

When there is indeed a need, be quick to use one or more of the tools required, but don't allow any of them to become a driving obsession. There are operations that have expended literally every available resource in setting up a highly sophisticated Kanban system, taking months and sometimes years to fully implement. In the meantime, the factory, as a whole, remained driven with the same basic operating procedures of old. Falling into this rut should be seriously avoided if the right kind of progress is to be made.

The end result of Kaizen activity can usually be placed in one of two categories: the type that is highly effective versus the type that is strongly disruptive. "Effective Kaizen" occurs when it serves to make lasting change for the better, the kind of change clearly recognized as such by management, shop floor operators, supervisors, and others. Kaizen that is truly effective becomes something extremely visible in terms of improvement and can be measured in inventory reduction, advancements in throughput, the removal of significant scrap and rework, and the enhancement of operator efficiency, to name a few of the more important.

"Disruptive Kaizen" on the other hand, is any form of Kaizen activity that leaves shop floor operators, supervisors, and others feeling frustrated, uncertain of its value, unsure of its intended goals, and generally confused about its merits. Unfortunately, there is as much disruptive Kaizen in modern business as that which is truly effective and it has nothing to do with the talent and ability of those involved. For the most part, what it has to do with is inadequate planning, execution, and follow-through.

It is extremely important to get the most out of each and every Kaizen effort conducted, because every effort that ends up with less than clearly positive results adds an ounce of lead to the anchor of skepticism. Preaching the right message is to no avail if management doesn't see the kind of results that will serve to motivate them to invest a lasting level of support in the effort. The same holds true if the workforce doesn't perceive it as something of real value to their jobs and the future.

Many Kaizen events end up being show-and-tell affairs, with no real strategy aimed at coupling the changes made to a master plan of implementation. This truly fits a "disruptive" Kaizen description. It's defined as disruptive because the employees involved are made to feel Lean is important and that the changes they have diligently worked to make is part and parcel of the revision of the entire factory over a period of time. When they begin to see the area revert back to the old ways—even slightly—it becomes an indicator that the company really isn't serious about Lean (which of course isn't always true) and that their work and efforts for the most part were wasted time and energy.

Good effective Kaizen efforts leave no doubt as to the value of the process and become the building blocks for pursuing further change in an aggressive manner. But far too often a company's Kaizen process begins to falter as time goes by. This doesn't occur because there wasn't a clear opportunity to make some very strong accomplishments. It happens because enthusiasm for the process, on the part of both management and the participants involved, slowly begins to wane. As things begin to waver, employees start to see the outgrowth of a "hybrid" system of production that incorporates some of both worlds (batch and Lean), with no apparent strategy to fully and completely change the way business has always been conducted.

One of the biggest enemies of the full insertion of Lean Manufacturing is that it's easy for it to be perceived as just another one of the countless initiatives and special programs that have come and gone over the years. Every company has them and for some organizations it's somewhat of a way of life. Although Lean efforts aren't squelched as being less important than any of the other programs or processes undertaken, Lean implementation commonly isn't elevated to a level of exceptional importance, with the understanding that without it becoming an integral part of any true progress made, the company faces a future of uncertainty.

Matter of "Misguided Pragmatism"

Some people refer to this as paradigms, but paradigms alone aren't the entire problem. Proclivities involve the tendency to behave in a particular manner or to like a particular thing. Pragmatism is striving to base judgment on practical solutions rather than the theoretical. Either one of these is harmless in itself when it comes to Lean Manufacturing, but combined as a mindset they can serve to establish serious roadblocks.

The issue comes to bear when leadership outwardly expresses (and essentially accepts) that change is needed, but has the proclivity to cling to the old or usual way of doing things. That on its own wouldn't necessarily establish a roadblock but coupled with strong feelings that the "old" or "usual" is the only practical or reliable way and you have what I like to refer to as "misguided pragmatism." When push comes to shove, anything other than the usual way is essentially perceived as being experimental in nature. This in turn sets the stage for slippage in implementation and all sorts of excuses for not aggressively pursuing needed change to some clearly established level of accomplishment. This is actually more subconsciously driven than specifically intended, but can be a force in opposition to making the kind of change a factory needs. American industry, on average, has to learn to recognize this particular flaw and manage to overcome any hurdle it may establish in making Lean a full reality. Otherwise a truly energetic thrust, aimed at making U.S. manufacturing world-competitive, will never take shape as it should.

What the mind-set of every employee (including management) has to come to be is:

It's a new world and we have to fully eliminate the practices that are slowly but persistently taking us down a familiar road toward a destiny of high uncertainty.

Once that hurdle is fully cleared, the only question becomes how to make it happen in the fastest and most effective manner.

Change can be forced, with the hope that once it's done the right steps were taken and nothing of importance was overlooked. But a much more workable approach is elevating the use of a well-known commodity in Kaizen to a serious focus by the workforce and from there allowing the process to work by getting out of the way of progress.

If it sounds like that infers management can sometimes inadvertently get in the way of needed change, that's exactly what it's meant to say. Not intentionally, of course, but because of operating practices that have served to feed misguided pragmatism. In making needed change, a conscious effort has to be made to remain receptive to new thoughts and ideas, and especially to new ways of doing things. Doing this sometimes means stopping long enough to think hard before reacting negatively to changes that might not initially come across as being practical or absolutely necessary. This is especially true when it comes to the matter of Kaizen.

Two Major Do's and Don'ts of Kaizen

In order to get the best out of Kaizen, it is important to start by understanding the "do's" and "don'ts" associated with a viable Kaizen process. A couple of the more important to mention on the front end are:

1. Kaizen should never be performed unless the maintenance function is completely aligned and fully capable of supporting the effort. This means having the necessary resources and materials available that allow change to be made in a fast and effective manner, at the participating team's discretion. What frequently happens, however, is that maintenance support becomes an afterthought and as a result the initial goals and objectives established on the front end of a Kaizen event end up being grossly underachieved.

2. Kaizen should never be performed unless the right people are involved. Involving the right people means establishing a cross-functional group of participants (both hourly and salaried), a few of whom have full decision-making powers without seeking approval from anyone. In most cases this means the operating manager of the area where the change is scheduled to take place, along with representatives from various support functions, such as purchasing and materials, quality control, scheduling, and so forth.

In my numerous consulting ventures, I cannot truly point to one Kaizen event where all the appropriate participants were scheduled to attend and participate. There were always plenty of excuses as to why not, and although the events usually turned out to be successful, this would not have happened if I had not taken it upon myself to literally demand the involvement of the right, required, people. How to make this happen without such an influence is one of the keys to making Kaizen a formidable competitive weapon.

Creating an environment for a truly effective Kaizen process centers on:

1. Understanding the different types of Kaizen and when and how to use them.
2. Developing a well-structured process for training and implementation
3. Driving Kaizen thinking through the rank and file.
4. Changing the principal role of the shop floor supervisor.
5. Providing advanced training for a select group of hourly employees.
6. Creating a high level of support and enthusiasm within the upper management ranks.

Evaluating and Rating a Company's Kaizen Efforts

To begin, it is important to take a look at how effective an operation's existing Kaizen process actually is. There are eight factors listed that should be rated: "Yes," "Somewhat," or "No." Given a conscious effort to be as factual as possible, this will help point out the strengths and weaknesses of an operation's Kaizen process in meeting what should be considered minimal expectations.

An accumulated score of 75 is extremely good and anyone achieving that level of accomplishment is well on his way to making Lean

1. A formal schedule for waste reduction activity has been developed and approved by top management, which details when Kaizen events will be conducted and what specific areas or production processes will be involved, along with the expressed purpose for the exercise, the ultimate goal to be achieved, and the estimated impact on the bottom line.

   Yes ☐ Somewhat ☐ No ☐

2. A member of the senior management staff (reporting to the plant manager) holds responsibility for Lean Manufacturing activities, including associated goals, objectives, and overall results. In addition, a full-time individual has been appointed and trained, as needed, to conduct and oversee Kaizen activities and overall Lean implementation.

   Yes ☐ Somewhat ☐ No ☐

3. A formal budget for Kaizen has been established and approved that covers anticipated expenses, including maintenance and standard event requirements for such things as completely laying out the area anew, developing new or revised fixturing, special handling devices, visual controls, and so on.

   Yes ☐ Somewhat ☐ No ☐

4. The plant manager holds biweekly Kaizen update meetings (at a minimum) that are attended by the full staff. At the meeting, goals, objectives, and results are reviewed and any problems discussed and fully resolved. In addition, the Kaizen schedule is revised as needed to meet new or unanticipated issues, related to the overall implementation of Lean and business in general.

   Yes ☐ Somewhat ☐ No ☐

5. At least one "High-Impact Kaizen" event is held annually, along with a minimum of two "Training and Implementation (TI) Kaizen" events monthly (roughly one every two weeks.) This standard relates to conducting a full event, which can range from two to three days for a TI event to one to two weeks for a high-impact event. *Note:* Specific information regarding a High-Impact and TI Kaizen event can be found in "Overview of the Various Types of Kaizen," in this chapter.

   Yes ☐ Somewhat ☐ No ☐

6. Some form of Kaizen is persistently used to address day-to-day production problems or issues responsible for creating downtime, scrap or rework, product quality issues, schedule deficiencies, and so on.

   Yes ☐ Somewhat ☐ No ☐

Exhibit 1.2   Kaizen evaluation form.

7. Plans are in place and efforts are being extended to train a vast majority of all employees, hourly and salaried, in the basics of Lean Manufacturing, including both classroom and hands-on shop floor Kaizen.

   Yes ☐   Somewhat ☐   No ☐

8. The importance of Kaizen has been thoroughly communicated to all employees and encouragement of the process is consistently highlighted and touted throughout the factory and the office arena.

   Yes ☐   Somewhat ☐   No ☐

Note: Score each "Yes" 10 points, each "Somewhat" 5 points, and each "No" 0 points.

Exhibit 1.2 (Continued)   Kaizen evaluation form.

Manufacturing a way of life in his facility. In an honest and unbiased evaluation, most factories in the United States would likely score 50 or below. But to be truly world-competitive, a score that approaches 70, at a minimum, is needed. Exhibit 1.2 speaks to the eight factors outlined in more detail.

Developing a Formal Schedule for Kaizen

Doing this requires a considerable amount of planning and forethought, including a look at the various types of Kaizen, along with when, where, and what the expressed purpose of the assigned effort will be. In the preparation and use of such a document an 18-month outline is a good start, with the understanding that the freedom exists to revise the last 6 months of the plan as conditions warrant. For every established Kaizen activity noted there should be a column headed "Purpose" where the reasoning for the effort is explained. This helps establish the overall logic behind the activity and assists in ensuring that nothing is taken for granted. There should also be a second column headed "Estimated Cost" where an effort is made to estimate the total expense involved, including required materials and the potential maintenance work involved.

Even though the initial plan prepared for and communicated to the workforce may only cover 12 to 18 months in duration, an overall master plan should be developed to include the timeframe estimated to change the entire factory to the full incorporation of Lean throughout. In doing so, the master plan could potentially cover a period of two to three years depending on conditions, the applied resources available, and other such matters.

The purpose of the master plan is twofold: to define the extent of the work involved and the approach that will be taken in making a total factory transition, and to estimate, to the best of one's ability, the cost of making such a transition and obtaining a buy-in from management to proceed. This document will undoubtedly change as time goes by, but every effort should be made to stick to it as outlined. Something every factory should guard against is changing the master plan to accommodate existing conditions. A far better approach is adapting conditions to support the plan. It all boils down to remaining fully dedicated to making the change required and doing so in the shortest period of time possible.

Without such planning there is absolutely nothing to ensure a solid management commitment regarding the extent and cost of such a venture. Conversely there is nothing for management to use to track the stated mission. Just as important is that the plan provide the Lean coordinator with a clear understanding of where and how to proceed. Such an understanding normally isn't the case for most Lean initiatives undertaken on American soil, which unfortunately leaves the overall application of Kaizen up to chance or how far someone of influence decides to pursue it.

Assigning a Qualified Full-Time Lean/Kaizen Coordinator

Coordinating a Lean/Kaizen initiative, resulting in the kind of change needed, requires a highly qualified individual dedicated full time to the effort. Anything less will simply not suffice in today's highly competitive environment. The person selected would ideally report to the plant manager or the individual seen as the ultimate decision maker for the factory. Even in the smaller operations the job should be full time until the factory has made an effective shift in its overall system of production.

There are essentially two alternatives to accomplishing this. The first is to take an existing employee who displays the ability to communicate effectively with others and provide that person with the essential training needed to assume the role. Ideally this would be someone who has had experience with Lean, otherwise the time required to bring them along would be extensive. The second alternative would be to hire an individual who has had some relatively strong experience in Lean Manufacturing, and has preferably led and taught others in the science of Kaizen. Either way, there has to be some reasonable confidence that the assigned individual has what it takes to make the kind of change required and is someone who will stand up as needed to see it accomplished. It isn't a job for the meek or the easily swayed and it requires strength of character and a willingness to challenge others, when and as the need arises.

For the company interested in pursuing the outlined process, care has to be taken to ensure the selected individual doesn't come armed with some seriously preconceived notions about Lean implementation, which could serve to hamper the strategy outlined for the advanced Lean implementation process. The individual needs to clearly understand that the roadmap for implementation has been firmly established by the company and the Lean coordinator's job is to see that it's carried out to the fullest. Most individuals with solid experience in Lean can easily relate to the components spelled out for ALIP, Progressive Kaizen, and so on, and can further understand the value of incorporating a WRAP initiative, once enough workforce training has been conducted.

Promoting and training internally is often seen as the proper thing to do for the role, considering the opportunity for advancement should exist and that the person stepping into the job would be a known commodity. But it isn't always the right thing to do. Lean implementation is a gigantic task if done properly; it best requires someone who has the proper knowledge and ability to direct a complete change to the existing system of production, and who then would move on to see that continuous improvement was made to the new system. It should not be considered a project-based role, which has a clear ending point, but rather one that requires a persistent and qualified influence on a long-term basis.

Establishing a Formal Budget for Kaizen

A budget for Kaizen should not be thrown into an overall training account or hidden within the confines of a standard budgeted line item. It should stand entirely on its own merits and be reviewed accordingly. The major reason is to keep the attention level high and to ensure Kaizen doesn't fall short of its intended quest. In the course of a standard budget review, when Kaizen activity becomes buried in another line item, there is absolutely no way of knowing for certain if it is being actively and aggressively pursued as intended.

The budget for Kaizen should represent the intent of a "Master Kaizen Plan" prepared by the Lean coordinator and approved by management. All associated expenses should be covered in the assigned budget; including employee training costs, maintenance expense, and the like. How to construct a master Kaizen plan is covered in Chapter 5 and is built around the parameters of ALIP. If ROI (Return On Investment) doesn't approach ten-fold or more, it's time for a sitdown for a serious discussion about the depth of Kaizen activity being pursued.

On the other hand, immediate cost savings should not be the principal measurement of success. As an example, one of the major benefits of a good Lean/Kaizen initiative is space savings, which doesn't pose an immediate payback. But as time goes by space savings provide the opportunity to lay out the entire factory anew and make room for added product or increased volume, without the expense of brick and mortar, which can be a substantial cost savings to any company.

Number and Type of Kaizen Events Conducted

A common deficiency that exists concerns the lack of conducting Kaizen events on a regular ongoing basis. There are many reasons involved for this discrepancy, such as other priorities, new initiatives, revised production schedules, and more. But the chief culprit is due to management not making certain that other facets of the business do not serve as stumbling blocks to fully implementing Lean.

Over a period of time, Kaizen events often become less and less important, which happens as a result of the effort providing inadequate results or an inability on the part of the factory to maintain the changes that have been made. When an operation reaches this point it's in danger of completely losing the momentum required for Lean implementation. Thus, all the more reason to ensure a formal schedule for Kaizen is prepared, approved, and tracked on a continuing basis, which would include the precise number of events that will be conducted.

Both the exact number and the type of events will vary from operation to operation depending on various conditions. But at least one "high-impact Kaizen" event should be held annually, along with at least two "Training and Implementation (TI)" Kaizen events monthly (see "Overview of the Various Types of Kaizen" in a following section of this chapter for more detail). This would mean at least 20 formal Kaizen events should be conducted annually, at a minimum, up until the time Lean Manufacturing has a firm foothold on the entire operation. In defining what a firm foothold means, it essentially boils down to a time when every member of the workforce has received some level of effective hands-on training in Lean.

Scope of Kaizen Training

The scope of Kaizen training should involve all salaried employees, as well as a large percentage of the hourly workforce. Ideally, each and every member of the workforce, including clerical positions, would be required to attend and participate in at least one TI Kaizen event. The idea, of course, is to provide everyone with firsthand knowledge and experience in the process. Doing this is especially necessary if Kaizen is driven down to the individual job level. More regarding this can be found in Chapter 2, under "Value of Inserting a WRAP Initiative."

Any manufacturing operation that is truly serious about Lean Manufacturing should insist that every employee be exposed to Kaizen training. This helps the entire workforce understand the importance of waste reduction activity and how the mechanics behind the process of Kaizen can apply to any job, whether it is on the shop floor or in the office arena.

Overview of Various Types of Kaizen

Properly defining Kaizen work cannot be lumped into one general category. In reality there are four distinct types of Kaizen. The first is <u>High-Impact Kaizen</u>, which is aimed at making dramatic improvements and solid inroads into revising the way production is conducted in a given area of the factory. Out of this activity will generally come the extensive training of mid- to high-level managers and supervisors in Lean Manufacturing, along with providing the experience of making hands-on change on the shop floor. Very often the end result is a showcase area that is representative of where the factory, as a whole, is headed in the future. The second type is <u>Training and Implementation Kaizen</u>, which is aimed at training the workforce over a span of time and making smaller but important changes on the shop floor. The third type is <u>Problem Resolution Kaizen</u>, which is directed at resolving recurring production problems and putting them to bed permanently. The last type is <u>Sustaining Kaizen</u>, which is used to make additional ongoing change to the initial improvements implemented in a high-impact and other types of formal Kaizen activity, along with ensuring that new equipment and production processing are installed with good Lean Manufacturing practices in mind.

The approach and technique for each type of Kaizen noted is different, because each represents different end goals and objectives. It's therefore important to plan and implement strategy accordingly.

High-Impact Kaizen is defined as making large sweeping change to an entire production area of the factory, normally involving a recognized shop floor department, such as a select final assembly line, a welding or brazing department, and the like. The training and implementation effort is extensive and aimed at entirely remethodizing and rearranging the area involved, setting in place the overriding principles of Lean Manufacturing, reducing space requirements, substantially changing flow, and dramatically reducing work-in-process inventory.

In some cases, the appropriate plan of action will call for fully decentralizing the department and placing equipment and operators at point-of-use. Performing this type of Kaizen requires a high level of participation from almost every support function, including quality assurance, production control, scheduling, purchasing, and even accounting, sales, and marketing under certain circumstances. This type of Kaizen is performed sparingly due to the time commitment and cost involved, but it's extremely important and knowing when and just how far to take it is vital to making Lean Manufacturing a full and absolute success. The specifics for this event can be found in Chapter 4 under the heading, "Conducting the Factory's First High-Impact Kaizen Event."

Training and Implementation Kaizen is a mini version of High-Impact Kaizen and is performed for the expressed purpose of providing knowledge to the entire workforce over an extended period of time. However, there is also a secondary objective called "implementation," which refers to making meaningful change on the shop floor. In the training and implementation Kaizen event an area of the factory is selected and change is carried through to completion or near completion. Ideally, the group reassembles at a later point to audit the changes made and to follow up on any work that was impossible to complete during the scope of the original event.

Training and Implementation Kaizen requires a solid commitment from management to ensure a majority of the workforce (both hourly and salaried) receives a minimum of 16 to 24 hours training in Lean Manufacturing. The task can be stretched over a period of time but the goal should be to have 75 to 80% of the workforce trained within a 12-month period of starting a Lean initiative. If an operation has been into Lean for greater than 12 months and hasn't as yet achieved that objective, they are running behind the timeframe that should be established for this particular objective. The specifics for training and implementation Kaizen can be found in Chapter 4, under the heading, "Getting the Most Out of Training and Implementation Kaizen."

Problem Resolution Kaizen is used to correct a situation that is seriously affecting throughput, quality, or the ability to achieve customer requirements. It can be used in areas where high impact and other types of Kaizen have been applied, along with those that have not as yet had Kaizen performed. The idea is to resolve a recurring production problem and to do so within the parameters of good Lean Manufacturing principles, so it remains an effective change as the factory makes the shift from batch to pull production.

Here, the emphasis is usually placed on an individual piece of production equipment or a group of like equipment. The Kaizen event is typically project-based and led by the plant's Lean coordinator, utilizing plant engineering personnel and hourly employees attached to the process. Again, the idea is to quickly and permanently resolve a problem within the parameters and guidelines of solid Lean Manufacturing practices. The specifics for this event can be found in Chapter 4, under the heading, "Driving the Use of Problem Resolution Kaizen."

Sustaining Kaizen is defined as making incremental improvements to an area that has had high-impact and other types of Kaizen previously performed. The sustaining Kaizen event is notably shorter in duration and principally involves personnel tied to the area involved (i.e., the production supervisor, various production employees, and select sustaining engineering personnel). On the other hand, it is good to include a number of fresh participants, as time and resources allow, who are given the opportunity to learn about Lean Manufacturing and how the process works. Such participants often bring a combination of fresh ideas to the table, because of not being influenced by the typical way of doing business.

Although considerable change is normally achieved during a high-impact event, there are usually items that cannot be fully completed over the course of that or any other Kaizen event. Sustaining Kaizen is the tool for completing change to its fullest. In such a sustaining Kaizen event, certain members of the original Kaizen team and a small number of those directly tied to the area come together for a two- to three-day session in order to fully complete various projects that were left unresolved. A good way to look at sustaining Kaizen is as an insurance policy to make certain that what was started with other forms of Kaizen activity is fully and completely accomplished. In doing so, opportunities for further improvement will usually surface. The specifics for this event can be found in Chapter 4, under "Understanding the Role and Scope of Sustaining Kaizen."

Most companies do not categorize Kaizen activity, but there is an important difference in the various types of Kaizen being performed that should be recognized, inasmuch as the approach, technique, and tactics applied will and should vary accordingly. Metaphorically, it's somewhat like participating on both sides of a football squad. Both the offensive and defensive sides of the game can be described as "playing football." But the devil is in the details and the strategy and tactics used to play football on offense are immensely different from playing defense. Viewing all forms of Kaizen as essentially one and the same is somewhat like approaching both the offensive and defensive sides of football with the same strategy in mind, which will almost guarantee that the ultimate objectives of Lean are never carried out to the fullest.

There are indeed special Kaizen events that should be planned and handled accordingly. High-Impact Kaizen would fit that category and a portion of Training and Implementation Kaizen would also apply. Sustaining Kaizen, on the other hand, should in no way be viewed as a special event, but rather an integral part of a plant's day-to-day activity. In addition, Problem Resolution Kaizen should be used to address and resolve the many production issues that typically arise in a factory trying to make a shift from a batch-driven system of production. One way to look at it is if some form of Kaizen activity isn't occurring each and every day in a factory, Lean definitely isn't being applied as it should.

Recapping, there are four distinct types of Kaizen that should be recognized, each requiring its own specific plans and strategies:

Type I: High-Impact Kaizen
Type II: Training and Implementation Kaizen
Type III: Problem Resolution Kaizen
Type IV: Sustaining Kaizen

Each of the various types of Kaizen noted works in unison with the others to funnel improvement into the foundation for the full insertion of Lean Manufacturing. No particular type is fundamentally more important than another. Each has its distinct purpose and role in the overall equation. However, under the best circumstances they would be introduced to a factory in the order noted.

High-Impact Kaizen is the best tool to introduce a factory to Lean Manufacturing. It shows the depth and extent of the change required and can serve as a showcase for each and every employee, visitor, and others to see. If done properly and used as an example, there will be few supervisors and production operators who will not see it as something positive, because even for those who have not as yet established a good understanding of the process, the showcase area will almost always leave the impression of being a substantially neater place to work.

For companies that have used some sort of High-Impact Kaizen, the biggest mistake made is not requiring supervisors of other areas in the factory to bring their workers for a tour and to make the point that the entire plant is scheduled for the same type of change. Doing this sets the stage for ensuring that everyone knows where the factory is headed in the future. But this has to be further reinforced with continuing communications and a plan of action to take the process to the next step.

A High-Impact Kaizen event is usually one to two weeks in duration and involves a cross-functional group of participants. Depending on various factors, the expense of such an event can be notable. However, if done right, the return on investment can be extremely significant.

Training and Implementation Kaizen should be started as soon as possible after the development of the showcase area noted. The objective should be to train as many employees as possible and make changes on the shop floor that are in keeping with an overall game plan to expose each and every area of the factory to the process. At some point it will be necessary to lay out the entire plant anew to support the implementation of Lean more readily across the entire production arena. In all likelihood this would be the first of two (possibly three) such changes in overall plant layout; which given an aggressive implementation schedule, would take place over an 18- to 24-month period. This is also where the greatest cost of implementation would be required, in the form of moving equipment and training employees. But again the money would be well spent and the return on investment should be noteworthy.

Problem Resolution Kaizen can most effectively be utilized after Training and Implementation Kaizen is actively in place: operators and others will have been trained in the basics and thus have a better understanding of Lean Manufacturing in general. The first objective of problem resolution Kaizen is to get down to the root cause of a problem. This requires a select group of people to go through a careful brainstorming session before taking action of any kind. Doing this avoids spending time, energy, and effort on fixes that do not fully resolve the matter.

There are cases where root cause is inherent to the design of the equipment, which can sometimes be expensive to resolve. When this occurs, it will usually take a top management decision to either live with the problem for a given period of time or move forward immediately with a capital appropriation. Either way, the problem will have been clearly identified and all the responsible parties alerted accordingly.

Sustaining Kaizen is again used in two manners. The first involves making further improvements, as needed, to changes made using other forms of Kaizen. The second use of sustaining Kaizen becomes the effort made to sustain the overall thrust to Lean Manufacturing, once a factory has made an entire shift to its system of production.

Progressive Kaizen Initiative

Coupling the various "types" of Kaizen under an all-encompassing process aimed at fully and effectively inserting Lean Manufacturing is the charter of "Progressive Kaizen." Figure 1.4 indicates the applied scope of each particular type of Kaizen event, the typical event duration, and the number of participants involved, along with the depth of change normally conducted.

Precisely What the Term "Event" Means

Some clarification could be warranted regarding the precise definition of how the term "event" applies, inasmuch as it's spoken to repeatedly throughout the content:

- A Kaizen event is a formally structured activity that takes a select group of participants away from their normal jobs for a specified period of time. That time can range from one to two days, to one to two weeks, depending on the type of Kaizen activity involved.
- A Kaizen event has two distinct purposes. The first is to train employees in the value of Lean Manufacturing principles and how to use the tools involved. The other is to make effective change to the production area or business process being addressed, which can be a single piece of production equipment, an entire line of equipment, a business activity such as order entry, or even an entire department or established production area of the factory. For the shop floor, the scope of change is

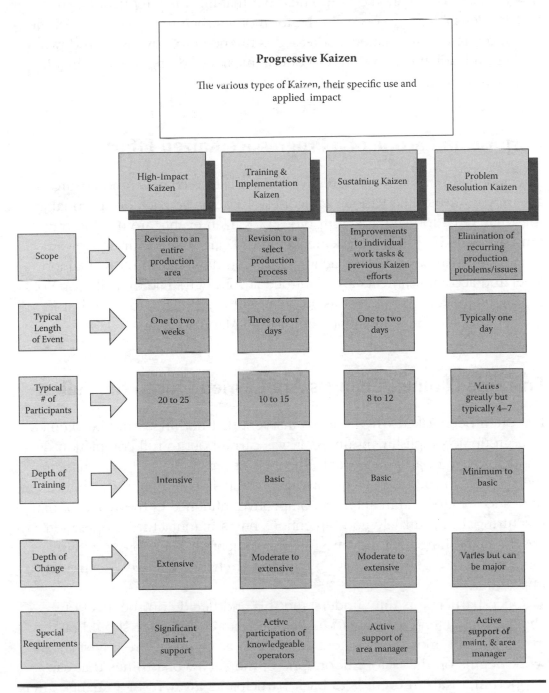

Figure 1.4   Progressive Kaizen.

essentially unlimited, depending on the number of participants involved and having the adequate support of the plant's maintenance function.

■ Special note: It should be pointed out that all waste reduction activity doesn't have to occur in the form of a structured event. One of the more important matters addressed is the need to drive Kaizen down to an individual job level and to offer a means of doing this in a highly effective manner.

Purpose and Scope of a Progressive Kaizen Effort

The assigned purpose of Progressive Kaizen is to recognize and ensure an appropriate use of the various categories of Kaizen activity, both formal and informal. On a formal basis, four types of Kaizen events are used to train employees and move the implementation of Lean Manufacturing forward. On an informal basis, Kaizen activity is driven down to an individual job level and accomplishments are rewarded accordingly. The overall scope of a progressive Kaizen initiative is aimed at utilizing Kaizen to its fullest in accelerating the full implementation of Lean Manufacturing.

Ensuring Planned Changes Are Carried Out to the Fullest

I've often been asked why I was so insistent that nothing could be claimed as a group accomplishment unless it was carried out to full completion during the course of a Kaizen event. The answer lies in the fact that changing the system of production as quickly as possible is paramount, if America hopes to achieve and maintain a competitive influence in the world of manufacturing. Unfortunately good intentions buy a manufacturing operation absolutely nothing until they are fully implemented. Therefore, establishing a sense of urgency in making the kind of change needed is extremely important.

Once participants fully understand this and decide on the level of change they intend to strive to make, the Lean coordinator should take their plan to senior management and obtain a buy-in before any serious work begins on the factory floor. One of the very worst things that can happen in a Kaizen event is to back participants away from a change after floor work has begun. It sends the wrong message to everyone involved. If any serious doubts linger regarding the change the team would like to make, management should take it upon itself to personally address the group and explain the reason why. A word of caution, however, is carefully to avoid delaying, minimizing, or completely stopping change that falls within the parameters of good Lean Manufacturing principles. If the situation boils down to a matter of not being able to afford the change being proposed, simply say so and work to build an understanding as to if and when it can be fully carried out. Otherwise strive as diligently as possible to accommodate the team's plans.

RELATED EXPERIENCE: In a two-week Kaizen event I was conducting for a company, we were almost done with the first week's training and ready to enter the weekend making change on the shop floor. I was informed midday on Friday that the maintenance function had encountered a serious problem with the plant's air compression and the crew that was scheduled to move and relocate equipment for the event was going to be cut in half. This left us with a serious problem. We could proceed as planned and allow what gains could be made over the second week or we could do something else. I suggested that the remaining portion of the event be delayed until we could arrange a time for me to return to the factory and pick up where we left off. This required me to adjust my schedule and for the company to pay the expense of extra airfare, but it was agreed if I was willing to return at a later date the company was more than willing to absorb the added travel expense involved. To accommodate a revisit of the training conducted during the first week, prior to starting change on the shop floor, I arranged to return three weeks later and conduct a needed "refresher" session on Saturday morning. When it was all said and done, the second week of change went exceptionally well and the full extent of the changes planned by the group was carried out to completion.

There are indeed times when it is better off to delay the full completion of an event rather than minimizing the change the team involved plans to make. As an example, for pressing business purposes it could be that one or a number of the key participants is required to leave an event. Instead of trying to manage and accomplish less without their participation, simply stop the event and pick up again at a time when they are free to participate. This is especially true during the initial training and planning phase of the event.

The goal of making every effort to see that planned changes during an event are fully and completely implemented should apply whether the services of an outside consultant are involved or the plant is conducting an event entirely on its own. This again goes back to building a sense of urgency in the need for change and striving to remove any obstacles that could potentially get in the way.

Production Manager's Role in a Kaizen Event

One of the biggest commitments of the factory's production manager, when it comes to Lean, is direct participation in a formal Kaizen activity. Because titles vary from company to company, the production manager is defined as the individual normally reporting directly to the plant manager, who holds the responsibility for managing and directing the day-to-day activities of the factory's production workforce. The following addresses a set of actions that would ideally pertain to this position:

- The production manager should personally attend the opening of any and all Kaizen events and say a few words in support of the effort.
- Over the course of the various types of Kaizen events conducted, the production manager should always attend the afternoon wrap-up sessions, where the teams report on results and discuss any particular issues that might arise. Although it is not absolutely essential, it is also beneficial for the production manager to occasionally drop by and observe some of the training going on. This shows a keen interest in the topics being covered and discussed.
- The production manager should further attend the closing session of an event, where participants make a presentation on the scope of the change and the results achieved. Here, the production manager would ideally be involved in passing out certificates of completion and offering congratulations to the group as a whole. Under normal event activities a plant tour is conducted after the closing presentation, which allows everyone in attendance to see the physical changes made. The production manager should always be present unless out of town on business and should ideally arrange to have a large portion of the reporting staff attend the group presentation and plant tour.

The important thing is for the production manager to show a keen interest and stay abreast of the training and changes going on. A Kaizen event that is properly conducted requires a great deal of work by the group involved and one of the best rewards is knowing their participation is clearly endorsed and appreciated by the highest level of factory management.

Key Summary Points

Elevating the Use and Effectiveness of Kaizen

It is extremely important to get the most out of each and every Kaizen effort conducted, because every effort that ends up with less than clearly positive results adds an ounce of lead to the anchor of skepticism. Preaching the right message is to no avail if management doesn't see the kind of results that will serve to motivate it to invest a continuing level of support in the effort. The same holds true if the workforce doesn't perceive it as something of real value to their jobs.

The Four Types of Progressive Kaizen

Properly defining Kaizen work cannot be lumped into one general category. In reality there are four distinct types of Kaizen: (1) High-Impact Kaizen, (2) Training and Implementation Kaizen, (3) Problem Resolution Kaizen, and (4) Sustaining Kaizen. The approach and technique used for each is different, inasmuch as each represents different end goals and objectives. It's therefore important to plan and implement strategy accordingly (see Figure 1.4).

Developing a Master Plan for Kaizen

The purpose of a master plan for Kaizen is twofold. The first is to define the extent of the work involved and the approach that will be used in making a total factory transition to Lean Manufacturing. The second is to estimate, to the best of one's ability, the cost of making such a transition and subsequently gain a buy-in from senior management (see "Developing a Formal Schedule for Kaizen").

Developing a Formal Budget for Kaizen

The established budget for Kaizen should not be thrown into an overall training account for the factory or hidden within the confines of a standard budgeted line item. It should stand entirely on its own merits and be reviewed accordingly. The major reason is to keep the attention level high and ensure that Kaizen doesn't fall short of its intended quest (see "Establishing a Formal Budget for Kaizen").

Applied Purpose of a Kaizen Event

A Kaizen event has two distinct purposes. One is to train employees in the value of Lean Manufacturing principles and how to use the tools employed. The other is to make effective change to the production area or business process being addressed; which can be a single piece of production equipment, an entire line of equipment, a business activity such as order entry, or even an entire department or established production area of the factory. For the shop floor, the scope of change is essentially unlimited depending on the number of participants involved and the adequate support of the plant's maintenance function (see "Precisely What the Term 'Event' Means").

Progressive Kaizen Initiative

Coupling the various "types" of Kaizen under an all-encompassing initiative, aimed at fully and effectively inserting Lean Manufacturing, is identified as "Progressive Kaizen." Figure 1.4 indicates the applied scope of each particular type of Kaizen event, the typical event duration, the number of participants involved, and the depth of change normally conducted.

Chapter 2

Addressing Key Roles and Supporting Tactics

No approach to the implementation of Lean Manufacturing can be duly successful without adequately addressing and revising the role and responsibilities of certain key players in the process. In addition, there has to be a well-thought out set of tactics that serve to support the effort. However, it all starts by establishing the proper frame of mind.

Clearing the Five-Inch Hazard

An interesting comparison to Lean implementation can be made to a comment Bobby Jones, the golfing great, reportedly said about the game. His comment was: "The toughest hazard to clear is the five inch space between the ears." The same logic could apply in many cases to Lean Manufacturing. In approaching the task of implementing Lean, the mind must be free of any lack of confidence. This is especially true of plant managers, who in turn have to see that those reporting to them do the same. Without this being accomplished on the front end of a Lean initiative, the chances of success are just as bad as the golfer whose mind is bombarded with doubt or anxiety about an upcoming shot.

Past experience to a large extent has to be disregarded and there has to be faith that the undertaking is unquestionably the right thing to do. The basic mind-set has to change from "What can I do to insert some level of Lean into the operation?" to "What can I do to make Lean a complete and unerring success?" Once that hazard is fully cleared, implementing the process can start to become a positive and rewarding experience, rather than a less than welcome challenge.

Enough can't be said about an absolute dedication to seeing that Lean Manufacturing is fully and effectively applied throughout an entire factory; and to further ensure this is done in the fastest manner possible. Otherwise what is currently left of manufacturing in the United States and any hopes to build it again to an adequate level of long-term competitiveness will only continue to dissipate. Without Lean, it's somewhat like trying to fight a raging forest fire with a squirt gun. One may be able to put out a small infinitesimal hot spot, but the huge fire of competition still blazes on unchallenged.

The reality every plant manager has to face is that the world of manufacturing has changed and every day of delay in correcting the practices of old will only serve to put a company another step behind the competition. But in addition, common thinking about roles and responsibilities has to change, if a full and effective shift to Lean is fully accomplished. The focus has to be greater than just getting added productivity out of employees. Added productivity will come naturally from having employees use their knowledge and abilities to the fullest extent. But this is not possible if plant leadership is not strongly confident and thusly motivated to meet the called-for challenge of the future.

Taking a Close Look at the Distribution of Change

As pointed out in the preceding chapter utilizing Kaizen to its fullest encompasses more than a single-minded process. Effectively using Kaizen calls for a series of established activities that have different purposes and lead to different results, all of which are aimed at fully and effectively inserting Lean Manufacturing. Figure 2.1 is a pie chart that outlines the typical distribution of accomplishments under a well-structured strategy for Lean, whether such activities are formally recognized as such or not.

Approximately 60% of the accomplishments will come from properly training the workforce and providing them a means to directly assist in making change in the factory, along with an effective use of a company's production engineering resource. Most manufacturing operations striving to implement Lean utilize both, from one extent to another. However, assuming the best from each area of application, this still leaves approximately 40% of the task that seldom receives proper attention.

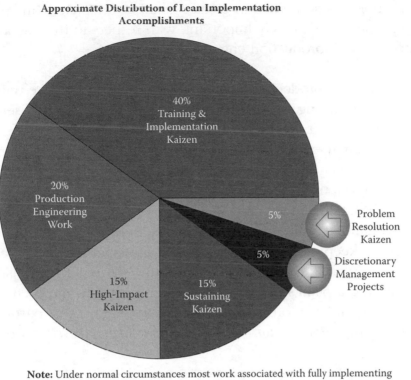

Approximate Distribution of Lean Implementation Accomplishments

- 40% Training & Implementation Kaizen
- 20% Production Engineering Work
- 15% High-Impact Kaizen
- 15% Sustaining Kaizen
- 5% Problem Resolution Kaizen
- 5% Discretionary Management Projects

Note: Under normal circumstances most work associated with fully implementing Lean will come from good production engineering practices, along with training the workforce and allowing them to directly participate in making change on the shop floor. However, unless Kaizen is used to its fullest—to include High Impact, Problem Resolution and Sustaining activity, along with various discretionary management projects aimed at enhancing the process—Lean Manufacturing will typically never be fully and successfully inserted.

Figure 2.1   Pie chart: Distribution of Lean accomplishments.

Unless Kaizen is used to its fullest, to include proper attention on high-impact, sustaining, and problem resolution Kaizen, along with various discretionary management initiatives aimed at enhancing the process, it is highly unlikely that Lean Manufacturing will ever be fully and successfully implemented.

Although the precise percentage in the distribution outlined will vary depending on numerous factors, what is shown closely approximates where the typical implementation of improvements can be expected. This serves to point to the fact that without a well-prepared strategy a company is missing the opportunity to make Kaizen a formidable competitive weapon.

The matter of appropriately utilizing production engineering cannot be overemphasized. Most companies simply haven't given enough attention to the role this particular resource should play in the overall implementation of

Lean. This subject is addressed in greater detail in Chapters 2 and 3, but the following points are extremely important, with respect to the role and commitment of a plant's production engineering (PE) staff:

- Each and every production engineer should have extensive training in Lean Manufacturing and should hold written objectives that serve to promote his direct involvement in the process.
- The production engineering manager should have a strong working relationship with the Lean Manufacturing coordinator and partner in bringing about the kind of change needed.
- The PE function should hold special objectives and conduct special activities aimed at incorporating Lean Manufacturing throughout the factory, in keeping with an established, management-approved implementation plan. Included in this is holding the chief responsibility of engineering a plant's key production equipment to more effectively support Lean. For more specific detail with regard this particular task, see *Lean Manufacturing; Implementation Strategies That Work.*

———————————————

Kaizen simply cannot be carried out to its fullest without a number of key positions playing a highly active role and without the formulation of tactics that serve to promote and drive the process forward. This starts with the plant manager.

Plant Manager's Role in Lean

It cannot be stressed strongly enough how important it is for the plant manager to be (or become) a strong proponent of Lean. Having personally worked and consulted with numerous factories across the United States and around the world, I can say that out of that total less than half of the plant managers involved expressed a seriously strong interest in making Lean a full reality. What is meant by a "full reality" is a clearly obvious sense of urgency in taking Lean to its ultimate level of achievement. It was clear some of them saw Lean as just another company initiative, among many that had come and gone over the years. In turn, they only did what was necessary to show a reasonable level of compliance. Others held extremely strong opposition to any change to the status quo and what was in keeping with the education and hands-on experience that had brought them to their current status.

But in fairness, plant managers typically aren't given the freedom to ignore standard performance measurements that serve to support this frame of mind. The active implementation of Lean Manufacturing doesn't come without a cost and, unfortunately, typical performance measurements often guide in an opposing direction. Therefore, unless a special fund is established for the effort, which is seldom the case, Lean and Kaizen will carry little weight in the overall scheme of doing business. That is precisely why so much effort has been put into striving to cost-justify changes made in support of Lean.

But any plant manager who believes his direct participation in the evolution to Lean essentially ends when he's hired a Lean coordinator and gone as far as communicating the importance to the workforce, is ill advised regarding the role a plant manager must play in the process. No less attention can be given to the implementation of Lean than any other important aspect of the business, and in most cases, a great deal more personal time and attention is required for Lean than other common initiatives and undertakings.

I can personally testify that the job of a plant manager can be a very time-consuming task, burdened with a wide variety of day-to-day issues, involving everything from public relations issues to operating costs, throughput, and satisfactorily meeting customer demand. If done right it's a tough assignment and one that deserves little criticism. On the other hand, one of the best ways to eliminate many of the distractions inherent to the job is for the plant manager to focus on seeing that a world-class production system is fully instated in the factory he is responsible for managing. Sometimes in doing so, he has to assume a "damn the torpedoes, full speed ahead" mentality. But for every plant manager who is comfortable assuming such a position, there are many others who simply won't take the risk. In the case of the latter, there has to be someone at a higher level who instigates and perpetuates the process or it likely will never come to fruition.

RELATED EXPERIENCE: We were heavily into making Lean-oriented change in a factory I was managing, when I was called on the carpet about expenses. It was pointed out that over the course of eight months I was $17,000 overbudget in training expenses. I explained that we simply hadn't budgeted enough for the aggressive workforce training we were conducting (although we didn't formally called the process Lean at the time). However, I went on to note that the return we were experiencing—in a reduction of work-in-process inventory and scrap and rework, along with substantial productivity improvements—more than justified the added training expense.

The comptroller in attendance quickly challenged my response, noting that although training expenditures were absolutely clear to everyone, fully contributing the cost improvements made to the efforts I was speaking to was "a seriously gray issue." What he was saying, in other words, was: "Prove it!"

I strived to do that, but continued to receive a tremendous amount of pressure on a monthly basis regarding being overbudget in training expense. That was until the president of the company happened to make an unexpected visit and tour of the factory. He went on to rave about the highly apparent changes that had been made since his last visit and his extreme pleasure in seeing it happen. Afterwards, I was never questioned again regarding the cost of the training expense involved.

The reason for relating that experience is to point out that regardless of proof, one should not expect every high-level official in a company to share a full understanding and appreciation for Lean or for the depth of accomplishment it is capable of achieving. There will be doubters and there will be surprise challenges along the way that the person pushing a Lean initiative will find frustrating and must be capable of addressing. Even under the best of circumstances, a willingness to extend one's self beyond a totally risk-free and comfortable position is almost assuredly required. Thus there is all the more reason for some appropriate planning and forethought before venturing into the effort.

Facing and accepting the need for a complete change to the existing system of production is usually the most difficult step to take. Unfortunately, there are plant managers who haven't fully brought themselves to this point. The lack of adequate senior leadership support for Lean varies greatly and is seeded with highly varying circumstances; but can be summed up in two basic categories.

The first category are the plant managers who give cursory support to the process, but play a highly inactive role in overall implementation. This type of plant managers are not personally involved in setting and directing plans and objectives for Lean, more or less leaving the depth of penetration up to the Lean coordinator and others. Although this category of plant managers don't typically create roadblocks, they do not go out of their way to strongly encourage Lean Manufacturing and the use of Kaizen in getting there. To them it is viewed as only one of any number of ways to enhance an operation, but something that isn't absolutely essential to the plant's overall success, or in other words, something they can take or leave.

The second category are the plant managers who makes it obvious that Lean Manufacturing takes a backseat to numerous other priorities, namely meeting scheduled forecasts and established production schedules, even if that schedule is aimed at building inventory that doesn't immediately satisfy customer needs and tends to create wastes that make the operation less than totally competitive. This category of plant managers will tend to work at delaying any real change to the status quo and the method of production they have worked with for years. Most often they have had Lean thrust upon them without their full agreement, perhaps as a result of a corporate-driven initiative, and will only do what is necessary to avoid being seen as defiant.

Most plant managers, however, are open-minded and willing to make change that serves to improve their operations. But something I can say with absolute certainty is that Lean is essentially doomed from the start if the plant manager is bound hard and fast to the practices of old and incapable of seeing the need for change and actively supporting it. Regardless of the measure of commitment taken, however, there are 10 actions that can be outlined as being characteristic of solidly Lean-oriented plant managers. While others could, of course, apply, the following highlight some of the more important.

Characteristics of Lean-Oriented Plant Managers

1. Play a highly active role in both establishing and following up on goals and objectives outlined for Lean Manufacturing, and Kaizen activity, in particular.

2. Make certain that all department managers and production supervisors carry written objectives that serve to enhance the full plantwide incorporation of Lean Manufacturing.

3. Ensure there is constant reinforcement in the form of written and verbal communications that point out results and the benefits achieved, along with the support the workforce should provide in making Lean Manufacturing a full and absolute success.

4. Ensure a formal budget for Kaizen is prepared, approved, and fully understood by all direct reports and that this is further reviewed with and understood by each of the staff's subordinates.

5. Attend the opening and closing of all Kaizen events, making positive comments about the process, extending congratulations, and, where appropriate, praise to participants for the results achieved. In the case of being absent from the factory or for an understandable inability to attend, the plant managers take the time to prepare a set of videos that can be used for the opening and closing sessions of each event.

6. Conduct regular tours of the factory with a number of the staff, for the expressed purpose of reviewing progress, auditing stated results, and making notes to share with the Lean coordinator regarding noticeable problem areas and where further opportunities for improvement potentially exist.

7. Make personal visits to the factory floor to speak directly with production employees and others about opportunities for change and about Lean Manufacturing in general, always taking the opportunity to note the need for everyone to actively support the process.

8. Express knowledge about the principles and techniques of Lean by directly questioning operators, production supervisors, and others when it becomes apparent that slippage has occurred to changes made. In absolutely no case would the plant managers knowingly walk by an obvious slippage in implementation without stopping to address the issue.

9. Hold at least one formal staff meeting a month centering on Lean Manufacturing; constructed to review measurements on progress, how training is going, all future plans and activities aimed at advancing Lean, and what can be done further at a management level to enhance the process.

10. Make certain that the Lean coordinator is a direct report and take the time to sit down at least once every week with the coordinator to discuss overall progress, any issues that might distract from getting the job done, and any opportunities that would serve to further advance the process.

The role of plant managers wishing to ensure a highly successful and productive Lean Manufacturing effort includes being both directly involved in associated planning activities and instrumental in creating a high level of enthusiasm among the workforce. This isn't to say that other company initiatives are not important, only that the majority of them are not quite as important as making a full and effective change to the system of production. In most cases Lean Manufacturing and the adequate utilization of Kaizen to get there should reign as the supreme objective and the plant manager is key to establishing and maintaining this way of thinking throughout the ranks.

In addition to the actions noted, there are some supportive organizational changes that the plant manager should strongly consider implementing. The following addresses two of the most important needed in support of Lean.

Lean Coordinator

One of most vital organizational issues required for a viable, energetic, and results-oriented Lean effort is to make the person selected as the Lean coordinator a staff-level employee, reporting directly to the plant manager. This means the Lean coordinator would be on par organizationally with the likes of the quality assurance manager, the materials manager, and others. This isn't always an easy thing to accomplish, inasmuch as Lean coordination isn't typically viewed as a high-level management position. This is because the individual usually heads up a very small staff of personnel (one to two, at best) and holds no special obligation to day-to-day production activities. The coordinator therefore carries no direct control over a sizable portion of the money required to run the factory, something most organizations take into consideration in setting staff-level positions. But old-fashioned thinking frankly has to change if Lean is expected to be accomplished in an aggressive and meaningful manner.

One manner of influencing the overall assigned responsibility and thusly the dollars controlled is to have the maintenance function report directly to the Lean coordinator. If the plant manager doesn't have the assigned authority to make such organizational change on his own, there are ways to approach getting there. One of them is to have an existing member of the staff assume oversight responsibility for Lean, along with his existing duties. This takes an enthusiastic staff manager who is well versed in Lean, who sees the benefits of the process and is willing to take on added responsibility. In such a case, the Lean coordinator would report directly to the selected staff manager and preferably dotted-line to the plant manager. As time goes by and change is made across the factory—improving the plant's ability to drastically lower inventory levels, reduce lead-times, and better service customers—it becomes easier to sway thinking into making the Lean coordinator a full-time staff member.

The fact is that a good Kaizen effort will normally outweigh the impact on the bottom line for which most staff positions have responsibility. One of the best examples for comparison is the purchasing manager position, which carries the responsibility to save multiple thousands of dollars each year through price negotiations. A good Lean Manufacturing/Kaizen coordinator can also save multiple thousands of dollars that are reflected in profit improvement for years on end thereafter.

Maintenance Manager

As mentioned, another important organizational change is to have the head of the plant's maintenance department report directly to the Lean coordinator. The principal reason for this is that good maintenance support is a critical component of any viable Lean Manufacturing initiative and is in fact vital to a strong results-oriented Kaizen effort. Having the head of maintenance report to the Lean coordinator takes any arguments and delays out of the equation. A further reason is that it takes maintenance projects that could be seen as highly important, but do nothing in support of Lean, and provides a strong voice in redirecting priorities. Taking this step means the Lean coordinator would preferably have some experience in maintenance or be capable of quickly learning the ropes. Should this not be the case, the plant manager has to decide how to overcome this particular shortcoming and make the organizational change at some point down the road. Regardless, it is an organization alignment that's needed and the sooner it's fully accomplished, the better a plant will be served.

F Alliance

In order to reach the active core of Lean implementation and gain the full benefits the plant manager (along with the Lean coordinator) should strive to adopt the "F Alliance" noted in Figure 2.2. Ideally, the plant

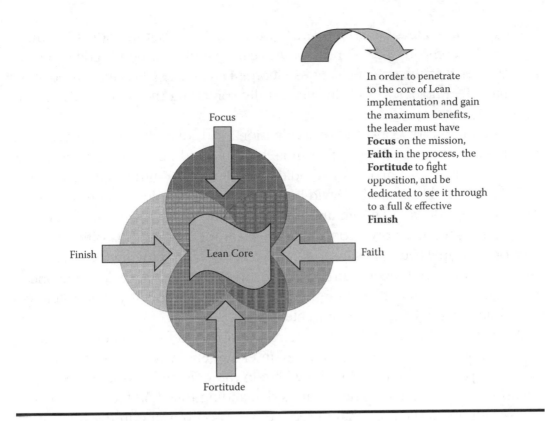

Figure 2.2   The F Alliance.

manager and Lean coordinator would be <u>Focused</u> on bringing about the kind of change needed to fully incorporate the principles of Lean across the entire factory. They would further have the <u>Fortitude</u> to make certain the process didn't stray off course, and would hold a strong <u>Faith</u> in the merits and value of the process. Last, but far from least, they would energetically strive to see it through to a complete and thorough <u>Finish</u>. Can a Lean initiative succeed without this level of conviction and involvement from the plant manager, in particular? Perhaps, but not close to the speed and effectiveness it can with a plant manager who adopts and expresses the values noted.

Lean-Oriented Company President

In most cases, a good Lean Manufacturing initiative will never get off to a start unless the company president sees a value in pursuing it and provides the initial push to get the process underway. I was fortunate to have worked for a highly Lean-oriented CEO, who I made note of in a recent book, *Lean*

Manufacturing; Implementation Strategies That Work. But in addressing the topic of the company president's role in Lean Manufacturing it's good to go a bit further into the specific actions George David, the CEO and president of United Technologies, took in successfully spreading the process across the entire corporation.

At the time, UTC consisted of six divisions with over 30,000 employees worldwide. All but one was a first in its industry and held such highly recognizable names as Carrier Air Conditioning, Pratt & Whitney, Otis Elevator, and Sikorski Helicopter. Mr. David became interested in learning more about the Toyota Production System and invited a group of ex-Toyota managers who had started their own consulting business for a visit to his office in Hartford, Connecticut. He subsequently went about employing the firm to provide a series of Kaizen training and "demonstrations" at various factories within UTC and came to see the importance of spreading Toyota's system of production throughout the entire enterprise.

Mr. David didn't hand the ultimate responsibility off to the next level and go about business as usual. He played an extremely active role in shaping an extensive plan of action and making certain the message was loud and clear regarding expectations; going as far as demanding that 1,500 top-level managers throughout the corporation receive hands-on training in the process. This meant taking time away from highly important jobs for a solid two weeks, in order to become an active participant in one of a series of special Kaizen events held at participating factories. It wasn't just an unusual commitment; it was a first of its kind and clearly demonstrated the importance, along with David's full intention of adapting Toyota's system of production to the highest degree possible in each and every factory within the realm of United Technologies.

I was asked to work in helping to develop a training manual for the venture and went on to spend four years traveling the world, conducting two-week Kaizen events for UTC manufacturing operations in the United States, the Far East, Europe, and South America. I can therefore testify to the magnitude of change made and the outstanding results achieved in plant inventory reduction, improved throughput, manufacturing lead-time, and the elimination of scrap and rework, among many other notable accomplishments. Such an achievement would simply not have been possible without the strong influence and personal commitment George David made, who later went on to be named one of America's leading CEOs.

Although it would be foolish to expect every company president or CEO to approach the matter with the same level of zeal George David displayed, I sincerely believe there has to be a renewed enthusiasm for the kind of change needed at the very highest level. Otherwise, America's future in manufacturing could seriously be at stake. In noting some of the important characteristics of a Lean-oriented company president, a few thoughts pertaining to the role follow.

The Lean-oriented company president would:

- Show an expressed personal enthusiasm for the process and the need for the company to follow the principles and procedures inherent to Lean, not only on the production floor but in all facets of work within the organization.
- Issue a directive that the company develop its own Lean Manufacturing training manual and that all manufacturing sites adopt and follow the basic outline prescribed.
- Make a personal commitment to visit as many factories as possible within the company, after a specified period of time, to review progress.
- Initiate an annual "President's Award" or something similar for the site that does the most effective job of implementation. In the case of smaller operations where the company president is located on-site, the award would go to the department that did the most effective job.

Fully incorporating Lean Manufacturing at an individual factory level and throughout a company as a whole requires the direct support and involvement from executive leadership, in order to ensure the process remains active and doesn't falter. Most companies and corporations that are sincere about Lean have a means in place to audit progress. But it's been my experience that many of these are aimed at rating one factory against another, rather than ensuring progress remains on track against a clearly prescribed implementation objective for the entire company.

The most important task for executive management as it applies to Lean is to establish a reasonable timeframe for full implementation and follow-up to see that it is carried out to the fullest. This will not be accomplished by making progress (or the lack thereof) some sort of contest, aimed at recognizing those who have done the most, inasmuch as being the absolute best in a given company can often fall far short of where the overall enterprise as a whole should set its sights. Instead, the executive directive and subsequent follow-up should be aimed at eliminating excuses and making certain that each and every factory involved makes solid irreversible progress toward full implementation. Anything less will simply not take the U.S. manufacturing sector far enough or fast enough to meet ever-growing competition, intent on establishing manufacturing superiority.

If that statement sounds a little strong, believe me it isn't. Manufacturing in the United States has already slipped significantly and the result over the past two decades has been the loss of countless jobs and the closure of thousands of factories. It is a very serious and sobering issue and one we won't overcome if we falter in the effort. At the core of that needed commitment is the strong support and the continuing encouragement of senior management in getting the job done.

Shop Floor Supervisor's Role in Kaizen

The shop floor supervisor holds a distinctive and important role in making Kaizen a formidable competitive weapon. In meeting the obligations of that role, he has to carry appropriate knowledge of the process and be active in its application. This means holding very specific goals and objectives, aimed at nurturing the process, such as having a certain percentage of his subordinates qualified as participants in the company's "Waste Reduction Activity Process" (WRAP), the key aspects of which are explained in an upcoming section of this chapter.

Being highly supportive also means going out of his way to ensure employees have the help they need in carrying out change for the better in their jobs, such as acquiring needed maintenance or engineering help in order to fully implement an idea an employee has in the advancement of Lean practices. Some of the more important aspects of Lean-oriented shop floor supervisors are:

1. They have gone through official orientation training on the specific requirements for the role and the responsibilities they hold in advancing Kaizen-related activity in their area of authority.
2. They practice encouragement of the process to subordinates and follow up to see that progress is being made throughout their area of shop floor responsibility.
3. They refuse to take "no" for an answer when it comes to the help needed from other functions such as maintenance and production engineering; and when necessary go to their respective bosses to solicit aid in getting the attention and support of others.

4. They consistently express enthusiasm for the process and push hard to make a complete change to their area of responsibility.
5. They have formal written objectives for Kaizen that carry a high weight factor for both performance evaluation and annual compensation.

With regard to the last item mentioned, an example of appropriate weighting is shown in Figure 2.3. The total of all the established objectives and their assigned weight factor should add up to 100%, 40% of which are Lean Manufacturing and Kaizen-related objectives. Why 40%? No specific reason, other than it's a large enough weight factor to grab one's attention and help to ensure a decent focus is maintained. Would something less than 40% be acceptable? Perhaps, but it definitely needs to carry enough weighted influence to ensure key players in the process give it the proper attention.

This should not lead one to believe that the purpose of applying a set of reasonably heavily weighted objectives, aimed at advancing Lean Manufacturing and the use of Kaizen, is to insinuate they are more important than other aspects of the business. All the categories typically noted in someone's management by objectives (MBOs) are important. But until a factory has made significant progress in fully changing its system of production, Lean Manufacturing and Kaizen objectives should carry an exceptional level of importance.

On the other hand, heavily weighting Lean and Kaizen objectives will not result in an individual's ability to successfully achieve them unless there is a good deal of understanding about the process and some extensive training. This is one of the reasons shop floor supervisors have to be some of the first and the best-trained individuals in a factory. How to go about this is covered in Chapter 4, under the heading "Training First-Line Supervisors." But in order to expound upon the importance of orienting shop floor supervisors to the task, it's noteworthy to relate a personal experience.

RELATED EXPERIENCE: In 2004, I was working with a company that had a formal MBO program that was used in assessing and rewarding performance. Under such a program an individual's annual merit increase is largely dependent on how well he or she goes about meeting written and approved objectives.

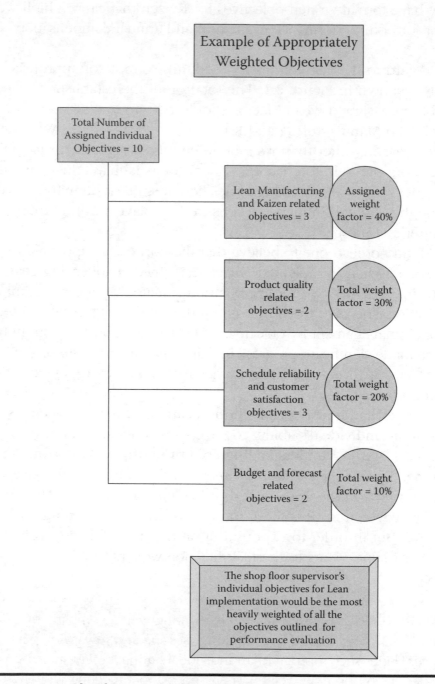

Figure 2.3   Example of appropriately weighted objectives.

After an initial audit of the factory's Lean Manufacturing process, I discovered that no manager or supervisor other than the assigned Lean coordinator carried any established objectives related to Lean. None! Upon questioning the plant manager, he informed me he had a bit of a problem in revising objectives because they had already been established for the year. He went on to say it would be the following year (some nine months down the road) before anything could be done. I left it at that for the time and went on to conduct a Kaizen session; even though it was apparent the production supervisors involved weren't that enthusiastic about the process and seemed to have their minds elsewhere. In my closing report I suggested to the plant manager that he do all within his power to see if MBOs couldn't be revised for some key individuals involved, to include at least some personal objectives aimed at enhancing the progression of Lean.

In a follow-up phone conversation some four months later, he informed me he'd gone about addressing the subject with his boss and had been able to get an agreement to revise formal objectives for a number of his people. He went on to add, perhaps not so surprisingly, that doing so definitely made a difference and that more had been accomplished since realigning objectives than had happened from the time Lean had been started in the factory. A couple of other interesting things he related were:

- He only had one supervisor who couldn't handle the responsibility and he ended up assigning the individual a different role where his expertise could best be utilized.
- The feedback he was getting from the shop floor was that the majority of employees were enjoying the changes being made and that union grievances were down overall.

Individual accomplishments to a very large degree turn out to be the things people feel are viewed as important by their direct leadership (i.e., their boss), in other words, the things they clearly understand as being fully expected of them. One of the most critical steps in directing the course of a Lean Manufacturing initiative is aligning clearly established objectives that serve to support the effort and leave absolutely no doubt regarding the expectations of management and the obligations an individual holds to those expectations.

Special Consideration of "Owner-Operators"

It is good to remember that Toyota didn't reach the status it holds in manufacturing excellence by making Kaizen a solely management run and operated process. It drove the mind-set and thinking down to the employee level and provided them the opportunity to make change for the better on their own.

We have to do no less if we hope to make manufacturing what it needs to be and successfully compete in today's extremely challenging environment. Unfortunately, there is far too little of this type of thinking in manufacturing and the only way we're going to change it is to do something different than is commonly practiced with Kaizen. But in addition, a company has to take a hard look at established labor classifications and make supportive adjustments.

Labor classifications can be a sore subject, inasmuch as it is often felt these are designed by the union to help the worker and not the company. To the average supervisor they stand in the way of effectively using the production workforce, as needed, in order to meet customer demand. I contend labor classifications are important and that the only reason a company has far too many classifications is because someone in management took the most convenient way out and allowed it to happen. I therefore find it hard to sympathize with a company who complains labor classifications are grown completely out of control. The answer is actually simple, although the task itself may not be. The answer is to sit down with the union and do something about it, the sooner the better! It won't always be easy, but it can be truly worthwhile.

I had a very high-level manager tell me he felt there should only be one labor classification and it should be called "worker," the idea being that any production employee could be used as needed. I pointed out to him although that would be nice with respect to making his job easier it wasn't necessarily the best way to go. He seemed a little startled at the statement and asked what I was trying to imply. My response essentially boiled down to this. If done properly, established labor classifications can be as much of an advantage as a disadvantage. The issue is clearly understanding where there's a need for a special classification and what elevates the responsibility beyond normal production work.

Far too often companies have given in a bit too easily to establishing a new classification for work that required no real training and where there was nothing particularly significant about the added responsibilities the operator was asked to assume. As a ground rule, there should be no added labor classification for something that doesn't require uniquely special skills and an expressed ability to pass a written exam or an on-the-job test of ability. Although a certain level of growth in experience could merit an increase in base pay, it in no way justifies a new labor classification. It's just that simple.

A very worthwhile consideration a company can make regarding classifications is the insertion of one known as "owner-operator." Becoming an owner-operator involves special training, which comes in the form of both problem resolution and sustaining Kaizen. The official qualification involves a period of basically uninterrupted classroom training, however, in order to ensure production requirements are met while training is being conducted, participants are given the first and last hour of the day to make certain that those serving as a replacement have the benefit of their input and guidance.

The classroom time for the three-day event is six hours a day. If a production issue arises, the supervisor has to essentially consider the participant absent from work for the period of time spent in class each day and strive to deal with the matter on his own.

In addition to the formal training given, once a year all fully vested owner-operators are required to take a six-hour refresher course. But an investment in owner-operators doesn't come without a cost. The work of the owner-operators requires a higher pay scale and special training, but can provide a very noteworthy payback. The return on investment comes in terms of greatly reducing the chance of machine breakdowns and successfully eliminating scrap and rework, along with the elimination of other wastes and inefficiencies common to the job. As pointed out later, a factory that doesn't aspire to owner-operators, or some similar classification, simply isn't getting the most out of Kaizen and potentially never will.

The specifics regarding this can be found in Chapter 4, under the heading "Training First-Line Supervisors," where the potential elevation of the position to "Lean Equipment Specialist" and the benefits involved are additionally discussed.

Value of Inserting a WRAP Initiative

A standard Kaizen event involves a select group of participants taking time away from their day-to-day jobs and working as a well-supervised team, under the direction of a qualified instructor. A waste reduction activity process (WRAP) takes Kaizen a step further by providing a means and an award for employees that apply Kaizen to their individual jobs.

WRAP encourages and reinforces the use of Kaizen with an incentive that pays a bonus for proven audited results. Taken to the ultimate would include constructive disciplinary action, when participants fail to respond in applying Kaizen to their job after appropriate training. The choice to take it to the ultimate is entirely up to the factory involved, but there are some distinct advantages in doing so that are addressed later in this section.

The term WRAP serves the process well, identifying it as a waste reduction process that essentially wraps Kaizen in a structure of strong encouragement, a high degree of employee participation, and award when achievements are successfully carried out. However, there are certain things that must be in place before undertaking a WRAP initiative. These include:

1. A highly supportive and dedicated management team
2. A willingness to pay a bonus for job improvements
3. A well-trained and enthusiastic group of shop floor supervisors
4. A highly knowledgeable Lean coordinator
5. A fully supportive human resource and accounting function
6. Cooperative union leadership (should one exist)

The goal of WRAP is to drive continuous improvement down to an employee level. The idea is to create a mind-set that waste reduction is an expected part of the job. For those who either cannot or will not comply after appropriate training, disciplinary action is taken. Such action should be aimed at striving to bring the employee along rather than rewarding failure with criticism and harsh measures. A positive approach would include additional training and direction as needed. But again, taking a WRAP initiative to this level requires floor supervisors and department managers who are well versed in their roles and responsibility to the process.

In the vast majority of cases, going as far as terminating an employee will not occur; although it can indeed happen if a company is truly serious about the process. In most cases what is being asked of an employee, after appropriate training, should be well within his means to accomplish. But in order for such a process to work, the single most important factor rests with well-trained shop floor supervisors or in the case of salaried employees, the respective department manager.

Managers and supervisors have to be trained and motivated to respond in obtaining the help employees need to carry out suggested ideas for improvement. For shop floor employees this usually centers on the direct participation of the plant's maintenance and production engineering functions. For salaried employees, it principally often boils down to the manager working with other department managers in bringing about change that reduces redundancy, paperwork, and other related wastes. But in both cases, employees are encouraged to make as much of the change as possible on their own.

If a labor union exists, expanding job expectations to this level can sometimes create a sizable hurdle. The bargaining chip is the matter of added remuneration, along with the fact that the company would not be asking employees to do more, without rewarding them accordingly. On the other hand, after appropriate training, the company should expect results and if an employee expresses an inability or unwillingness to participate, the company should have every right to take action, which could include as a last resort, termination.

Far too often training is given to employees and little to nothing is achieved other than the ability to say training was indeed conducted. I once had a discussion with a production manager who spent a considerable amount of time showing me all the various training his workers had received. The truth was I felt that little of the training was in line with what employees should be receiving and went on to ask if he thought his workforce was "better" than the average production worker found elsewhere. He pondered the question a moment before responding, "To be truthful, I can't really say for sure. But I think they're among the best."

As the conversation proceeded, I was able to learn his turnover rate was relatively high, which forced me to question if it was due to their moving on to better and higher-paying jobs or because of other factors. This spurred his interest and what we collectively came to discover was that the feedback typically given human resources in exit interviews, almost always boiled down to job-related dissatisfaction.

As we learned more, it became a prime example of a company spending money, time, and effort on training that essentially resulted in little to nothing in return. But again, even with the best training, floor supervisors and department managers have to play a highly supportive role in a WRAP process. In order to make WRAP work correctly, the role of department managers and area supervisors has to change from being highly directional in nature to becoming much more motivational and supportive of employees. This isn't something that can or will happen overnight and will take time. But if approached in the right manner a change in role responsibility can indeed take place.

The bonus paid for waste reduction activity at an individual employee level should be something meaningful. It doesn't always have to be purely monetary and can involve such things as earned time off and the like. However, bonuses should be individualized as much as possible. The idea is to reward personal involvement and results. Precisely what amounts to a change that earns a bonus has to be at the discretion of the company involved. A reminder, however, is not all good improvements end up immediately saving the company money. Some ideas will serve to enhance the overall progress of changing the existing system of production and should not be overlooked as a successful accomplishment.

It is sometimes helpful to consider a flat bonus for qualified improvements ($50 to $100, for example) rather than striving to apply a complicated formula based on a number of extensive reporting parameters. Keeping it simple and straightforward, by taking any complicated formulas out of the equation, is one of the keys to making a WRAP initiative work to both the company's and the employee's advantage.

One very positive aspect of the term "waste reduction activity" is that it clearly points out the scope and purpose of the work involved. It isn't something to make someone's job easier, although it may indeed do that. It isn't work to make someone feel better, although that may also happen. It is work designed to make the company less wasteful and just a little better today than it was yesterday.

In Chapter 4 more is provided regarding the mechanics of setting up WRAP under "Implementing a WRAP Initiative," along with the appropriate planning and follow-through required in order to make it a success. However, it goes without being said that an entirely different mind-set is warranted, along with a focus that hasn't commonly been placed on job performance at an individual employee level. Can a company implement an advanced application of Kaizen as outlined without a WRAP initiative? It probably can. But it will not gain the substantial benefits of making Kaizen a "daily activity" and fully accomplish the value of continuous improvement at an individual job level.

Tactics for Getting the Best Results Out of Kaizen

The tactics involved in advancing Lean and getting the best results from Kaizen start with a plan regarding how to go about the task. Figure 2.4 outlines a series of progressive steps to establish a firm foundation for Lean implementation most effectively. Charts 2.1 through 2.3 outline the steps involved in more detail and provide an implementation timeframe for each phase of the process, under what could be considered a relatively aggressive application of Lean.

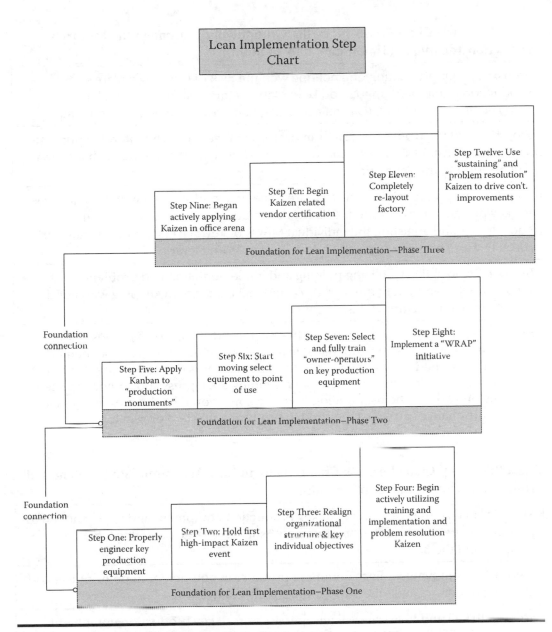

Figure 2.4   Lean implementation step chart.

How to Use the Step Charts

For those just starting a Lean initiative the steps noted could be followed precisely as outlined. But because many factories already have a Lean initiative underway, the starting point would then be different. Assume there's a factory that has been into Lean for a year or so. Some notable changes have been made, but have been slow in coming and rather sparse in application.

Chart 2.1   Step Chart Phase 1: Lay the Groundwork to Change the Factory's Production Technique (Timeframe: 8 to 10 months)

Step #1: Assign production engineering with the task of bringing a plant's key production equipment up to good Lean Manufacturing standards (see *"Lean Manufacturing; Implementation Strategies That Work"* for more detail on this step).
Step #2: Hold the plant's first "High-Impact Kaizen" session. The intended purpose should be to establish an area showcasing both the extent and type of change that will be conducted plantwide.
Step #3: Realign organizational structure as needed and see that written objectives for key players are established. This would include assigning a full-time Lean coordinator and adjusting the individual MBOs of key players in order to support the active and ongoing implementation of Lean Manufacturing.
Step #4: Begin actively utilizing training and implementation and problem resolution Kaizen to train the workforce and make change in keeping with good Lean Manufacturing principles.

Special Note: The aim should be to see that each production area is exposed to at least one TI Kaizen session and that at least one active pull zone is implemented in each assigned area of the factory. "Assigned area" is defined as an established production department, such as the welding area, press shop, wiring subassembly, and so on. The user pulling from the zone would be the next assigned area in the chain of activities required to produce a finished product.

Chart 2.2   Step Chart Phase 2: Change Flow to Best Accommodate Lean and Pull Production (Timeframe: 6 to 8 months)

Step #5: Finalize the engineering of key production equipment and apply Kanban to production areas considered "monuments," such as a large paint line, coating process, and the like.
Step #6: Start moving a select amount of production equipment to point-of-use and allowing operators and others to learn from the change.
Step #7: Select and fully train owner-operators on all key production equipment.
Step #8. Implement a formal Waste Reduction Activity Process and associated Lean Manufacturing incentive program.

Some machines have had Poka-Yoke and SMED applied and Kanban has been established at various locations in the factory. However, there is clearly room for improvement. The steps outlined can still be used. As an example, the plant could decide to start at Step #1 inasmuch as this has not been accomplished. However, because Lean isn't new to the factory, a decision

Chart 2.3 Step Chart Phase 3: Fully Implement Lean Manufacturing Throughout (Timeframe: 9 to 12 months)

Step #9: Actively drive Kaizen into the office functions.
Step #10: Build supportive Kaizen-related change into vendor certification.
Step #11: Completely lay out the factory anew. Plan and progressively make the entire new layout principally over weekends or during established periods of shutdown for vacations, and the like.
Special Note: Fully incorporating the new layout should result in gaining space in the factory that can be used for purposes of adding new business in the future, along with the necessary production equipment required.
Step #12: Use sustaining and problem resolution Kaizen to further continuous improvement.

could be made to skip Step #2 and proceed to Step #3. The point is to use the charts to stay a proper course and make certain that key elements of the process are not overlooked. The important thing is to guard against changing the sequence of the steps involved. There is good reasoning and experience behind them and care should be taken before rearranging the sequence of activities outlined.

With regard to the phased timeframes noted, following the steps precisely as outlined would normally take a factory of any size somewhere between 24 to 30 months (roughly 2 to 2½ years) to completely and thoroughly implement. But compared to the typical progress made with Lean, a start to finish of 24 months is a gross improvement.

This chapter has served to point out that all Kaizen activity is not the same and should not be looked upon and approached as such. It further indicated the ideal characteristics of the highest levels of management, along with the roles a number of other key players should undertake and went on to cover the principal benefits associated with implementing a waste reduction activity process.

In order to make Kaizen all it can be, the general thinking within an organization has to change, it has to be adequately supported from a management standpoint, and it has to incorporate methods beyond performing the sheer mechanics of Kaizen. One might ask if you don't have the kind of management characteristics outlined, is a good Kaizen effort essentially doomed? The answer to that is not entirely, although it does require someone of influence who expends an effort to preach the right message and who works hard at getting senior management's support for a change in course.

Unfortunately without the insertion of many of the things addressed, which are covered in even greater detail as we move along, it's highly unlikely that any grand accomplishment will be made with Kaizen, and even less likely that a full and complete change will occur. But at some distinct point (which no one can predict for certain) if we fail to make a satisfactory change on a substantially large scale in the United States, we will move beyond the point of no return. However, we still have time if we approach the matter with the right frame of mind and are willing to seriously challenge the status quo.

Key Summary Points

- The matter of appropriately utilizing production engineering cannot be overemphasized. Most companies simply haven't given enough attention to the role this particular resource should play in the implementation of Lean Manufacturing (see "Taking a Close Look at the Distribution of Change").
- The plant manager must make certain that all production managers and supervisors, along with all needed support functions, carry written objectives that serve to enhance the achievement of the stated mission of Kaizen and the full implementation of Lean Manufacturing (see "Characteristics of Lean-Oriented Plant Managers").
- Having the head of the maintenance department report to the Lean coordinator takes any arguments and delays out of the equation. A further reason is that it takes maintenance projects that could be seen as important to the current leadership of maintenance (but which do nothing to support a viable Kaizen effort) and provides a strong voice in redirecting priorities (see "Maintenance Manager").
- Toyota didn't reach the status it holds in manufacturing excellence by making Kaizen a solely management-run and operated process. It drove the mind-set and thinking down to the employee level and provided employees the opportunity to make change for the better on their own. We have to do no less if we hope to make manufacturing what it needs to be to compete in today's challenging environment (see "Special Consideration of Owner-Operators").

- A standard Kaizen event involves a select group of participants taking time away from their day-to-day jobs and working as a well-supervised team, under the direction of a qualified instructor. A waste reduction activity process takes Kaizen a step further by providing a means and an incentive for employees to apply it to their individual jobs (see "Value of Inserting a WRAP Initiative").

- In order to make Kaizen all it can be, the general thinking within an organization has to change: it has to be adequately supported from a management standpoint, and has to incorporate methods beyond performing the sheer mechanics of Kaizen. One might ask if you don't have the kind of management characteristics noted if Kaizen is doomed. The answer is not entirely, although it does require someone of influence who expends the effort to preach the right message and who works hard at getting senior management's support for a change in course (see "Tactics for Getting the Best Results Out of Kaizen").

Chapter 3

Avoiding the Typical Pitfalls

As should be obvious at this point, Kaizen isn't something that can be left entirely to its own devices. It has to be coupled with a detailed plan of action that spells out the specific "type" of Kaizen that will be utilized, at what point in Lean implementation it will be used, and specifically where the effort will be directed and why. This requires the Lean coordinator, the plant manager, and potentially others to sit down and give serious thought as to how the system of production will be fully changed, giving consideration to the entire factory, from receiving to shipping. Anything less is basically a haphazard approach to Lean that likely will not take the factory where it needs to go.

The finished and approved plan has to be further reviewed and followed up on by senior management on a seriously active basis. It has to be something focused on daily and reported on and reviewed just as frequently as to how well production is going or how budgets and forecasts are being maintained. How to go about this task is addressed in Chapter 5, under "Constructing a Master Kaizen Plan."

We haven't as yet learned to be as proficient with Kaizen or the planning required that is called for by the change itself. That isn't intended as a criticism of typical Kaizen activities, but rather as an observation as to where U.S. manufacturing currently stands and what we must do differently to get the absolute best results out of the process.

In conjunction with a sound plan for implementation and the associated Kaizen activity involved, there are some typical pitfalls that should be kept in mind and avoided. We can start by looking at outside influences on the process.

Allowing Outside Assistance to Cloud a Path to Success

As a consultant in the field of Lean Manufacturing I respect the importance a qualified individual can lend the effort, especially on the front end of a Lean initiative or when significant problems arise with implementation. On the other hand, having also served as a plant manager I fully understand the difficulty associated with justifying an extraordinary expense that doesn't end up resolving an immediate production need. This points to the fact that precisely how and when consulting services are warranted, along with what the consulting activity is aimed at accomplishing, should be carefully evaluated.

There will generally be occasions where the assistance of a knowledgeable consultant is needed. However, the buyer should beware. Literally hundreds, if not thousands, of "Lean consultants" have emerged over the past decade, as Lean has grown in industry awareness and acceptance. Along with this has come a somewhat confusing mix of tactics and advice. It's therefore vital to know precisely what one is getting in the bargain when soliciting outside services to assist with the implementation of Lean.

There are numerous ways of going about selecting a good consultant. One is basing the selection on a recommendation from a trusted source who has used a particular service and was pleased with the outcome. But regardless of who is selected to discuss the possibility of using his or her service, there are some steps that can be taken to help the decision-making process.

Outside of the common items generally discussed, such as background, applied expertise in the field, and the like, the following are some pointers that serve to tell much about the candidates:

1. Ask them to accompany you on a plant tour and point out to you where they see various deficiencies and opportunities for improvement. Any Lean consultant who can't identify a number of improvement opportunities during a quick walk-through of a conventional manufacturing operation simply isn't qualified. You should end the interview as quickly as possible and look elsewhere.
2. Make a point to quiz them about the key accomplishments they've personally been involved with in managing or directing. Any good consultant will usually provide a list of clients with whom he or the company he represents has worked. But if he can't easily discuss improvements he personally participated in leading and speak intelligently to some very significant accomplishments, he simply isn't the kind of Lean consultant who should be considered.

3. Ask them to share a vision of what a fully Lean-oriented factory would represent, both visually and operationally. In that expressed vision there should be a description of a factory that has fully moved its system of production from conventional manufacturing practices to world-class approach. Among other things, the envisioned factory would include the following:
   - Extensive visual controls throughout
   - A fully incorporated factorywide pull system of production
   - The elimination of scrap and rework
   - Single-minute setup and changeover of equipment
   - Inspection and correction devices built in at the source of operation
   - A well-identified place for everything and everything in its place
   - U-Cell equipment and processing layout arrangements
   - Point-of-use manufacturing applications
   - A clear and evident use of SMED, Poka-Yoke, and TPM

   If the candidate's vision doesn't closely approximate this, it is highly unlikely that he is the kind of consultant that should be considered.

4. Ask them to provide a "Lean Manufacturing Plant Assessment" without charge, prior to obtaining an agreement for use of their services (other than the company covering the expense of any necessary travel, lodging, and transportation). Any Lean consultant who isn't willing to do this either has more business than he can handle or doesn't feel he could pass the test. Asking for this should be viewed as an insurance policy, of sorts, which will go a long way in pointing out the qualifications of the consultant and the services he can potentially render, spelled out in a summary report on the findings and recommendations outlined by the consultant involved.

The important thing is to get a good reading of the consultant's knowledge and ability to effectively address the needs of the factory in a manner that serves to build teamwork and understanding. Another thing to remember when using the services of an outside consultant is to guard against taking actions based entirely on his feelings about how things should be done. They can and should give specific guidance and input. But it's always good to remember that no matter how qualified a consultant may be, he is an outsider looking in and someone who isn't fully familiar with the interworkings of a given business or production process.

The best and most effective change will always come from the workforce itself. The consultant's job should not be to direct specific change, but rather to bring something to the effort that is truly needed in elevating both the awareness and the ability of the workforce. One of the hardest things for most Lean consultants is learning how to "back away" enough to encourage participants to use their own brainpower in making change for the better.

RELATED EXPERIENCE: I've always striven to guard against telling a factory precisely what changes should be made. This wasn't because I didn't recognize where many of the greatest opportunities rested. It was because I felt strongly it was more important to pass on the kind of training and know-how that led employees to see the opportunities and take a genuine interest in making change for the better. My opinion on that has never swayed.

In an event I was conducting, the team involved was struggling with precisely how to lay out an area anew. I was called aside by the plant manager and asked how I would go about it. My response was that although I could obviously insert my opinion about how the work area should precisely be laid out and the team would likely jump at using the input, doing so would essentially make it my layout and not theirs. I went on to assure him not to worry; I would coach the team as needed and help them guard against doing something that fell outside the parameters of good Lean Manufacturing principles.

The plant manager went along although there was a level of discomfort expressed. In the end, however, the team implemented a layout that was admirably better than anticipated and the factory went on to achieve some extremely good accomplishments. The question could be was the layout what I would have done on my own, and the answer to that is, not entirely. But the point to be made is that a much better layout was accomplished and Lean Manufacturing principles and techniques were strongly applied.

As most Lean enthusiasts know, in reality there is no such thing as a perfect layout, just a "better layout" where opportunities for improvement will always exist. That is why Lean in the larger sense is a never-ending process (something to which Toyota will testify) and why a strong ongoing Kaizen effort is absolutely essential. Both the depth and effectiveness of change are extremely important, however, just as vital is the learning experience it provides the workforce. In fact, to a large degree, the latter serves a more important and lasting purpose. Equipment and production processing come and go, but the applied knowledge of the workforce remains a resource on which to build. Therefore, actively involving the workforce in the process and providing them with the know-how is critical to making Lean a full and absolute success.

A truly effective Kaizen effort must be a strong collaboration between management and employees, which essentially says to everyone involved:

> We are in this thing together and will work very closely in bringing about the kind of change that serves to make the company more competitive and everyone's job more secure. In keeping with that mission, there will be no changes made without appropriate input and representation from those who end up inheriting the results.

Something that most manufacturing managers may not like to hear, but which is extremely important to a Lean effort, is that it's better to allow participants to make a mistake and come to see how to correct it, than to push them for absolute perfection.

What can go a long way in seriously cutting down on mistakes is to encourage participants to keep the four guiding principles in mind (workplace organization, uninterrupted flow, insignificant changeover, and error-free processing). By teaching participants these principles and requiring a substantial focus on them, it becomes as simple as asking the question, "Where does the change you would like to make fit with one or more of the guiding principles?" If the team can't readily point out where the principles apply, it's time to ask them to give the proposal more thought. Doing this provides the team involved with sound reasoning for asking them to rethink a proposal and goes a long way in avoiding the impression that any work and effort they put forth was simply "shot down" by the powers above.

Misstep of Excluding Office Functions

A major misstep manufacturing operations tend to make is not actively involving office functions and salaried employees in the Kaizen process. Although select salaried employees are sometimes asked to participate in helping improve the shop floor, the office itself has been virtually exempt from the process in many organizations. This happens because Kaizen is generally seen as an activity aimed at improving the production process and not the work performed in the office.

Although it is true enough that some Kaizen has been aimed at various office processes, for the most part office employees have been left out of the process, or perhaps better said, haven't been invited to actively participate in improving their own jobs. The very same guiding principles noted for change on the production floor apply to change in the office arena, especially workplace organization and error-free processing.

First and foremost in getting the ball rolling is recognizing and accepting that it is vitally important to involve salaried workers in the Kaizen process and from there to put some form of action plan in place to make it happen. Office Kaizen, as it is usually referred to, is somewhat more difficult to both organize and conduct than shop floor Kaizen. This is because there are usually few written instructions and very little standardization associated with common ongoing activities, as opposed to shop floor operations where things are generally spelled out in a precise step-by-step manner.

Using the process of Kaizen should be driven down to the individual job level to the greatest extent possible, across the entire enterprise. This is best achieved by giving each and every salaried employee training and establishing a platform that goes about encouraging them to apply Kaizen actively to their individual assignments. One manner of elevating this, as previously mentioned, is putting WRAP in place, which provides an incentive to making effective change. The amount and type of award varies depending on how a company decides to approach the matter of a bonus, but accepted improvements in the office would most often involve things such as:

- Clearly reducing paperwork or redundancy of effort
- Shortening the time span for a repetitive business process, such as order entry
- Improving the efficiency of one's job, thus creating the time to take on additional work and responsibilities

Under the best circumstances, a company would have a Lean coordinator who focused on training and making change to the production floor, along with a qualified assistant who focused time and effort on managing the company's WRAP initiative and working to advance Kaizen in the office functions. To provide a clearer picture regarding how such an approach would typically work, the following is noted:

> In a fictional company we simply call Manufacturing, Tim is the assigned Lean coordinator. He has a strong background in manufacturing and firsthand experience in Kaizen, and is recognized as having an advanced understanding of Lean and how to apply it. Working for Tim is Ellen, who comes without a solid background in Lean, but has been nurtured and coached by Tim. As part of her initial training Ellen was sent to a number of outside training seminars in Lean and Kaizen. Functionally, Tim spends most of his time conducting shop floor related Kaizen. Ellen spends her time coordinating the company's WRAP initiative and conducting Kaizen events for those in the office. Both Tim and Ellen work together as a team in organizing and facilitating the far less frequent but extremely important High-Impact Kaizen events, which normally take place once or twice a year.

A Special Word about Lean and a Company's Financial Arm

Most companies do not apply the amount of resources noted in the above example to their Kaizen initiative. In fact, many companies have an assigned Lean coordinator who typically carries the duties of the role, along with other production-sustaining responsibilities. This could be an industrial engineer, for example, who is not only responsible for conducting Kaizen training (on an occasional basis at best) but also holds a production engineering sustaining role in the factory.

Other examples could be given, but in order to make Kaizen all it can be, the process needs to be elevated into the office functions just as actively as it's pursued on the shop floor. Doing this in the proper manner requires an adequate resource base for coordination, which would ideally involve two to three well-trained and thoroughly qualified individuals at a minimum. working full time on the effort. When I've passed this particular suggestion on to various operations, I've often been told that the plant simply couldn't afford to make the added investment required. My response was always, "If you're really serious about Lean, you can't afford not to."

It has always amazed me how companies manage to justify what amounts to an inadequate resource base for Lean. The same people would never have the purchasing function, for example, run by someone on a part-time basis. On the other hand, they often see directing something as large and all encompassing as fully changing a plant's system of production as capable of being handled in just that manner. The end result is usually one person held responsible for all the planning, training, and coordination involved. It simply doesn't add up and I contend when this occurs management either fails to understand the extent and importance of the task involved or really isn't all that serious about Lean.

If Lean implementation is to be recognized and accepted for what it actually means in importance to the future, the financial arm of the company has to take care in viewing it as just another special expense. They have to be led to accept it as something absolutely critical to the long-term financial stability of the company. Unfortunately, achieving this level of awareness and acceptance hasn't happened on a large scale in U.S.-based industry, and it's something America's business leaders hold the chief responsibility in ensuring that it happens.

That aside, however, changing the system of production requires salaried support functions that understand the importance of improving their normal activities in order to support a full-fledged thrust to Lean. This frankly cannot be done unless each and every employee is given appropriate training in Lean, along with establishing a process that encourages them to use the principles of Kaizen in their day-to-day work.

Allowing Kaizen Accomplishments to Deteriorate

One of the most damaging things that can happen with any Kaizen effort is to allow accomplishments made to erode. This sends absolutely the wrong message. I have seen repeated cases of letting Kaizen accomplishments slowly deteriorate as time goes by. In the simplest reasoning as to why, it usually centers on letting other initiatives and the like overpower the progress of Lean. This often boils down to the workforce allowing attention to the measurements that drive the existing system of production take precedence over inserting and maintaining a much more competitive approach to manufacturing, and until the general mind-set of both management and employees alike becomes acutely directed at the task, things typically aren't going to change.

RELATED EXPERIENCE: I was working with a company that hired me to conduct a two-week Kaizen session aimed at developing a showcase area for the factory. What was meant by "showcase" was an area that incorporated all the key aspects of Lean Manufacturing that could be used to indicate to employees, visitors, and others where the factory was headed in the future.

The event itself went extremely well. Tremendous change was made. Workplace organization was taken to the ultimate, general product flow was completely revised, space was reduced, and excellent visual controls abounded throughout. The first true pull zone for the factory was instituted and it appeared the plant was on its way to driving the same type of change through the entire factory.

I came to find out later, however, that the pull-production concept never developed as intended and over a period of time, visual controls and other changes that were made gradually began to deteriorate. For reasons of confidentiality I can't get into the specific influences involved, but it's another example of taking a step in the right direction and then letting other issues get in the way of progress. Intentions were good and the need for change was firmly recognized, but the process of fully implementing Lean fell by the wayside due to "other pressing circumstances."

It would be nice if every manufacturing operation in the United States had a Taiichi Ohno (the recognized father of the Toyota production system) who held the power, leverage, and know-how to not only insist, but forcefully demand the kind of change required. It isn't of course realistic to expect this level of zeal from every plant manager or from those otherwise in charge of making the highest level decisions for a factory. As a result, the duty typically falls on an individual in a recognized coordination position, who, it is hoped, is willing to stand up and fight for not only maintaining the changes made, but persistently moving things forward with additional improvements.

To a large extent when a plant manager takes on a complete change to a factory that goes entirely against the grain of how things have always been done, it isn't something that can be approached in a tentative or less than aggressive fashion. Someone has to be adamant about the need for change, fully committed to seeing it through, and unwilling to take no for an answer. Under the very best scenario that person would be the plant manager. But regardless of who accepts the challenge, leading a complete change to the existing system of production isn't a job for the meek or ill prepared.

I can personally testify to the fact there is indeed the chance of seriously alienating some members of the workforce, up to the point of their doing everything within their power to cast as much doubt and despair as possible on the effort, including the person seen as leading that effort. Sometimes various outspoken employees who have been around an operation for years on end will see the change being conducted as a direct threat to their job security and will make a point of persistently criticizing the effort. There really isn't much that can be done to avert this, other than conducting appropriate communications as to the need and staying aware of treating everyone's feelings about the matter as fairly as possible.

But for those taking on the challenge of making necessary change and ensuring that it remains intact, my advice would be to push implementation in a fervent and energetic manner. Do the job that needs to be done, never fearing you may inadvertently make as many enemies as friends along the way because, as dark as things may appear from time to time, you will eventually be recognized for your efforts and rewarded accordingly. That much I can say with almost absolute assurance.

A practice the plant manager can use to ensure that the erosion of accomplishments doesn't occur is a "weekly Lean tour." In performing the tour, which had no set time or schedule for the purpose of making certain the floor didn't go out of its way to prepare things for the visit, I always asked the members of my staff who could break loose to accompany me and also took along the Lean coordinator.

During the "tour" as it came to be known, we reviewed any recent work that had been done (which assumes some form of Kaizen is performed every week) and chose an area where change had been made in order to make a closer examination. In the area selected we looked at what had been documented as improvements, comparing these to what was actually going on, on the shop floor, making certain they were still intact. We then went about discussing where further improvements could potentially be made, speaking with operators and others as required. If any slippage had occurred it was addressed with the assigned production supervisor, who followed up within three days with a written report indicating what had been done to bring the area back, at a minimum, to its original level of accomplishment. This served to tell the workforce that management was extremely serious about Lean Manufacturing, along with moving the process forward as rapidly and effectively as possible.

Although I cannot say that a Lean effort is doomed if the plant manager doesn't do something similar, I can say with absolute certainty that some form of adequate management audit has to occur, in order to ensure the erosion of accomplishments doesn't happen and to keep the attention level focused on Lean. It has been my experience that if the plant manager doesn't leave the distinct impression that he or she is solidly behind the effort, the odds are high that Lean will be viewed by most of the workforce as a less than important priority.

RELATED EXPERIENCE: I make a practice of providing a free plant assessment for a factory interested in an unbiased opinion about how they are doing with the insertion of Lean Manufacturing. In one such venture I was making a walk-through with the Lean coordinator.

At a series of metal cutting machines he was showing me the visual controls that had been put in place, which were very well done and gave the impression that good Lean practices were being utilized. Before leaving the area I asked him about a series of locked cabinets arranged at each machine. He informed me they were used to keep the necessary tools and components used for changing the equipment over. For a number of reasons, I've always disliked the thought of locked cabinets and asked if we could take a look inside. What we found both highly surprised him and confirmed my opinion about the practice.

The cabinet we observed was correctly labeled for the tools and components required, but some of the needed tools were missing or not in their assigned location. In addition, we found that the tool cabinets were being used to store various personal items.

After a search of the general work area the missing tools could not be found and it was discovered that a visual aid cross-reference had not been updated. The Lean coordinator immediately addressed the matter with those involved and made it clear he wanted a written plan of action that provided assurance the situation was fully corrected and never happened again. He went on to take the time to explain to the operators that although they would have probably been able to put their hands on the tools needed, the effort put forth in a search for them was nothing but wasted time and a distraction to efficiently meeting customer demand.

When he was finished, he apologized for the delay and we returned to the walk-through. As we moved along, I made the point to assure him if he'd done anything less than he'd stopped and taken the time for, I would have been seriously disappointed.

It could be said that this sort of thing could happen from time to time in any Lean effort. After all, there's always the issue of ongoing changes in personnel, written procedures, and so forth. What was pointed out, however, is actually an indictment of conventional manufacturing, which until it's been thoroughly revised will always include a lack of appropriate discipline. The important thing when slippage is discovered is to immediately correct the situation and leave a clear message that any reoccurrence will simply not be tolerated, which was exactly what was done.

Failure to Communicate the Full Extent and Scope of Kaizen

Something that ties directly to the preceding subject is a failure to communicate the full extent and scope of Kaizen to everyone involved. It is extremely important on the front end to let the workforce know:

- The vital importance of fully changing the existing system of production
- The approximate timing involved in making a full plant transition
- How far the process must go in terms of individual job responsibility

These are not always easy matters to discuss and doing so requires very thoughtfully deciding how to approach the subject. In *Lean Manufacturing: Implementation Strategies That Work* under "Getting the Message Over to the Troops," a complete speech is outlined, as an example of the kinds of things that should be addressed in the initial communications to the workforce. No one message, however, no matter how good it may be, will suffice as having properly communicated the full extent and scope of Kaizen. The first message has to be reinforced on no less than a quarterly basis, at least for a time.

It's important to remember what the workforce is being asked is to stop doing much of what they've been taught to do in the past and approach their job in an entirely different manner. Getting this across is complicated by the fact that it isn't something that's going to happen overnight. At best it's a relatively extended process: 18 to 24 months under a fairly aggressive application of Lean. It is somewhat like telling everyone, "I have a new job for you, but most of you will have to wait and see precisely what that's going to be." What most of the workforce is going to see is a rather extended transition in getting everyone fully trained and properly aligned. As a result, elevating and keeping the message going is one of the more important aspects of a good Lean initiative.

The key is to start the ball rolling and diligently pound the point home for as long as the need is there. This level of communication typically isn't achieved in most operations, but it's truly needed if Kaizen is elevated to the position it should hold. One manner of clearly expressing the purpose of Kaizen and keeping it in front of everyone is a visual displayed in numerous areas of the factory, which can be used to represent both the value and need of Lean.

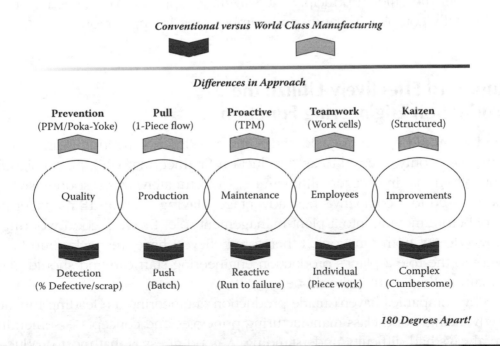

Figure 3.1   Conventional versus world-class manufacturing.

Figure 3.1 is only one of a number that can help with design. This particular chart may not be what is chosen in getting the word out and keeping a constant reminder in front of everyone; however, it denotes the differences that exist between conventional manufacturing and a world-class approach, Those differences are expressed in terms of quality, production techniques, equipment maintenance, the manner in which employees are expected to work, and how improvements to the operation are typically made.

For example, quality under a Lean Manufacturing approach is focused on prevention, utilizing parts per million as a measurement and Poka-Yoke as a tool for correcting and preventing quality issues. The conventional manufacturing approach is 180 degrees in opposition and focuses on detection, using percentage defective as a measurement and scrap and rework as a common solution. In a nutshell, the other major differences noted are:

- Production is "pulled" rather than "pushed" through the factory.
- Maintenance is "proactive" rather than "reactive" (using TPM versus running equipment to failure).
- Employees work as a team rather than having a strict focus on individual output.
- Last, but not least, the process of improvement is structured change under the guiding principles of Kaizen, as opposed to change that is usually complex in nature and cumbersome in application.

Failure to Effectively Utilize the Production Engineering Function

The Industrial Engineering (IE) and Manufacturing Engineering (ME) functions, commonly referred to as "Production Engineering" (PE), play a significant role in making Kaizen all it can be. Unfortunately, this resource isn't always used to a company's best advantage. Having been an IE manager for years before moving into a plant management role, I have a good working knowledge of both functions. I therefore believe I have the background to speak to the role a plant's production engineering staff can and should play in making Kaizen all it should be.

Why companies haven't made production engineering the leading function in advancing world-class manufacturing principles and concepts is something I've always had difficulty understanding. A good guess is that most production engineering functions have been downsized over the past two decades, which has left little time for anything other than sustaining work, involving such things as developing and maintaining process sheets, performing methods analysis and work measurement, establishing labor standards for product cost purposes, ordering and installing equipment, and the like.

I truly believe there's a need for companies to seriously reassess their production engineering function and more actively involve them in the Kaizen process. I further believe colleges and universities should put a much stronger emphasis on the educational aspects of Lean Manufacturing in order for graduates to come to industry with the fundamental knowledge needed to help advance the process.

In *Lean Manufacturing; Implementation Strategies That Work,* Chapter 5, "Choosing and Aligning the Engineering Staff," I addressed using production engineering to initiate Lean Manufacturing in a factory, with the principal focus on engineering a plant's key production equipment to be in tune with fully supporting an active Lean initiative.

Industrial engineering has long been viewed as a profession aimed at improving methods and efficiency. Manufacturing engineering is typically viewed as a profession that specializes in equipment and production processing development. Who then is better qualified to assume a lead role in Kaizen? The answer is obviously no one. But I have repeatedly seen production engineers grossly underutilized, with little to no real ownership in the process.

There are three key areas where production engineering becomes a vital player. The first is with a qualified application of Poka-Yoke, where efforts are made to install improvements to equipment, fixturing, and the like, aimed at avoiding common production errors and building quality in at point of use. The second area is in the advanced application of SMED, where setup on key equipment is cut to single minutes, at a minimum. Most of the expertise for SMED and Poka-Yoke will rest with a company's manufacturing engineers, but a third and extremely important area of expertise required is with "Standard Work" (methods and work measurement), which not only establishes where opportunities exist for productivity improvement, but sets the basics for both methodizing and standardizing repetitive production work.

Many manufacturing operations have been led to believe that almost any employee can be trained to adequately perform Poka-Yoke, SMED, and work measurement. This is far from actually being the case. Performing Poka-Yoke and SMED in an appropriate manner requires someone that is reasonably educated and skilled in equipment processing and design. Work measurement in turn requires someone that has an educational background in methods and time analysis. I've had companies that were strongly into Lean take a bone of contention with that position. But the fact of the matter is that Standard Work by design ignores performance rating for skill and effort, assuming the individual being studied is a fully qualified employee who aspires to giving a "normal" effort.

If there is one area where we've been somewhat misled or misinformed, it's in the area of work measurement or what is referred to in the Toyota production system as Standard Work. In my visits to Japan, I toured numerous factories that aspired to Toyota's system and witnessed excellent applications throughout. But not once did I see anyone performing work measurement other than qualified production engineers.

Having personally conducted literally thousands of time and motion studies during the early stages of my career, I can assure anyone that the feeling any employee can be adequately trained in a short period of time to perform work measurement simply isn't an assumption a company can afford to make. Standard Work teaches that throwing out various "highs" and "lows" in the times recorded suffices in overcoming any question regarding skill and effort. This again is an assumption that simply isn't valid. Regardless of how many individual readings are taken and how many of the readings are accepted or retracted, if an operator isn't appropriately skilled at the job or extends more or less effort than what truly represents a normal pace, the time derived from that study (used for establishing cost standards and other purposes) will be wrong. It's just that simple.

This doesn't mean hourly supervisors, production workers, and others can't be trained to take the data developed and utilize combination work sheets, percent loading charts, and other established tools of analysis to make improvements. They most definitely can. However, the work of gathering such data best fits a well-qualified industrial engineer.

Figure 3.2 points out where the conveyance of production engineering's talent should be directed. Best put, a plant should cautiously avoid the pitfall of not fully and effectively utilizing the production engineering staff in advancing Lean Manufacturing. But an operation's ability to effectively involve them depends on two key factors:

■ The function's overall work load and available resources
■ A company's willingness to acquire additional engineering talent, as needed, in order to actively support Kaizen and make a full thrust with Lean Manufacturing

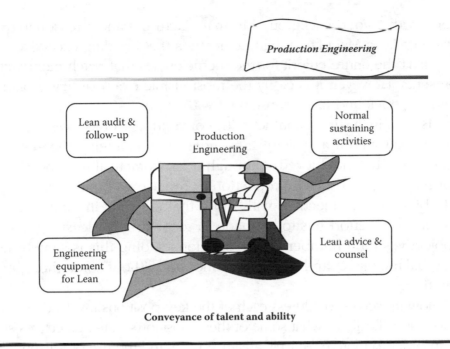

Figure 3.2   Production engineering—Conveyance of talent and ability.

Failure to Restructure the Stated Objectives of Key Players

Another flaw with many Kaizen efforts is the oversight of failing to restructure the formally written and approved objectives of a number of key players in the process. The positions that are absolutely critical to the effort are:

- The production manager
- The floor supervisor
- The production engineer

Most often the positions noted hold formal objectives that are essentially in conflict with the full incorporation of good Lean Manufacturing principles. Standard objectives normally involve things such as striving to increase operator output in order to theoretically "absorb" direct labor costs. There are many other such commonly stated objectives that apply. But when a company through its formal structured objectives inadvertently or otherwise encourages building excessive inventory, it is touting the things that go directly against what is hoped to be achieved with Lean Manufacturing.

Anyone with a good background in manufacturing and a reasonable knowledge of Lean Manufacturing understands that building excessive inventory isn't the entire culprit for the inefficiencies that batch manufacturing represents. However, it is easily the most visible example that a plant is on the wrong track and has a substantial way to go.

There is usually a direct conflict with conventional objectives that have to be dealt with if any significant progress in changing the system of production is to be achieved. Although it isn't practical to expect a firm that has long used batch production to replace its typical measurements and individual objectives with those that are fully in tune with Lean Manufacturing, an effort has to be made to reconstruct at least some of the stated objectives for a number of key positions. Doing this is sometimes a delicate balancing act and one that cannot go without some thoughtful consideration.

The following serves to address each of the key positions involved and speak more specifically to what some of these position's stated objectives should include, in order to more effectively advance Lean and the use of Kaizen.

Production Manager's Stated Objectives

The production manager is the individual in charge of all shop floor activity, aimed at ensuring schedule reliability, quality, and delivery. It's a big job and one that usually requires a forceful personality. This can be a good attribute in helping to aggressively push a Lean initiative forward, given that the production manager not only has a knowledgeable respect for Lean, but a few reasonably weighted objectives that will help to guide his or her efforts. Two very good related objectives to consider for this particular position are:

1. Ensuring an agreed-upon percentage of the hourly workforce attends a Kaizen training and implementation event and receives certification.
2. Ensuring an agreed-upon percentage of the plant's key production equipment is made available for a full application of SMED, Poka-Yoke, and TPM.

For a company that is serious-minded about Lean Manufacturing, such objectives should carry a combined weight factor that is no less than 40% of the total.

RELATED EXPERIENCE: In a conversation I was having with a plant manager on this subject, who very strongly supported Lean, he proceeded to make it clear that he felt a combined weight factor of 40% was simply too high. The conversation went something like this: "The plant as a whole is still operating under a batch system of production and this will continue until we end up making the kind of change we need across the entire factory. But in all fairness, I don't see how I can reasonably base almost half of how our people are measured for a potential merit increase on such a small percentage of what's actually going on in the factory."

"I believe you're missing my point," I replied. "I'm not recommending that you amend everyone's objectives. I'm suggesting you consider including some reasonably high-weighted objectives for a number of key players. Accountability has to start somewhere if you're going to make a success of it, and you need to keep in mind if they do things right in meeting a set of Lean objectives, it can only enhance the other facets of their job responsibility."

"I understand we need to get there as rapidly as possible," he responded. "But in the meantime we still have to run the business in keeping with the way we're measured and expected to perform. I simply have a fundamental problem with the weight factor you're suggesting."

"So what would you be comfortable with?" I proceeded to ask.

"I think 20% would be more appropriate and would certainly be challenging enough for the circumstances," he replied.

I told him that would be a start in the right direction, but asked him to do me a favor: "I'd like you to get an understanding with the people involved on a couple of key objectives aimed at advancing Lean and then allow them to suggest what they think the combined weight factor of those should be. From there you can decide what you're comfortable with. Who knows? They might surprise you."

He agreed and two weeks later I checked back to see how things were going. He related that in a conversation with some of his key people about the combined weight factor for a set of Lean objectives, that he'd apparently done a good job of selling the importance to everyone; going on to add they had ended up agreeing on 40%. "I didn't start out by jamming 40% down their throat," he noted. "We just ended up there as the discussion proceeded."

"So what was your major selling point?" I asked out of curiosity.

"That potentially all our jobs were at stake if we failed to focus some proper attention on making the kind of change needed," he replied, before adding with a chuckle: "Heck, you would have thought it was you talking rather than me."

The lesson that can be taken from that experience is not to let preconceived notions drive decisions pertaining to an appropriate environment for change. The change needed requires moving things along in a reasonably urgent manner and doing this calls for certain individuals to assume a strong supportive role, which likely will not happen without clearly defining some meaningful and challenging obligations to the process.

Shop Floor Supervisor's Stated Objectives

The shop floor supervisor is the production employee's direct link to management. The supervisor's actions therefore create a perception in employees' minds about management in general. That perception of course can be either good or bad depending on the factors that drive the supervisor's conduct and interaction with subordinates. The individual serving in this role typically has one primary mission whether it is formalized or not. That mission is to fully meet assigned production schedules and operate within the guidelines of an established budget, with the manpower given to accomplish the task.

One of my early jobs in manufacturing was working as a production foreman in a school and office furniture factory. I will never forget it, because it gave me a sincere appreciation and a relatively good understanding of the role and responsibilities of the job. I was 26 years old at the time and had much to learn about manufacturing. But I was somewhat shocked at the lack of actual authority the role carried. The people I was in charge of supervising were hired by the personnel department, for the most part void of any real input on my part. The work to be accomplished was established by a production schedule over which I had little to no control. Any and all serious personnel issues had to be directed to human resources, thus any disciplinary decisions boiled down to something passed down to me to enforce. At the time I saw the role as little more than a mouthpiece for higher authority and now, some 40 years later, I still have to say that feeling hasn't changed substantially.

American manufacturing needs to provide shop floor supervisors with a higher level of positive influence over the subordinates who directly report to them. Given that much of what a production foreman does is carry out a mission that has been handed down, Kaizen opens the door to providing the kind of influence I'm speaking about, the kind that can serve to strengthen the relationship between the supervisor and those who report to him.

To provide that kind of influence, however, there has to be some sort of added incentive attached to the results that come from Kaizen work, such as that outlined earlier for a WRAP initiative. Otherwise, you are not only asking the production worker to do his normal job, but to go further in assisting the company to become "world class," void of any reason other than conscientiousness to do so. All I can say to that is good luck!

Two excellent objectives for the production supervisor are:

- Ensuring a set percentage of the hourly employees reporting to the supervisor implement Kaizen-related improvements to their jobs
- Ensuring something is done in departmental communications to advance Kaizen, such as seeing that a Lean control board for employees is set up and fully utilized

Production Engineer's Stated Objectives

The production engineer is usually an industrial or manufacturing engineer assigned to perform sustaining work for a factory. This typically involves the development of routing sheets, bills of material, work measurement, and so forth. But as long as there is a Lean Manufacturing initiative in place and a company hopes to successfully advance it, there are some key objectives that production engineers should maintain. A couple of the more important are:

1. Ensuring there is a plan and process in place to bring a plant's key equipment up to speed, through an advanced application of SMED, Poka-Yoke, Standard Work, and TPM
2. Playing a highly active role in advancing Kaizen activities and lending assistance and counsel to others in the process, as needed

Doing this means production engineers have to be some of the best-trained individuals in Lean Manufacturing and specifically in a thorough application of Kaizen across the entire factory.

Error of Putting Lean in a Stand-By Mode

The last, but certainly not least, important pitfall to avoid is putting Lean Manufacturing on hold because of pressing financial matters or other factors. As mentioned, fully incorporating Lean isn't free. There is a cost involved including the training of employees, the expense of moving and rearranging equipment and facilities, and the cost associated with bringing in outside talent to assist as needed, along with the salaries of those dedicated both full and part time to the effort. Lean is a discretionary expense that is often chosen as one that can be placed on a stand-by mode if needed in order to slash operating expense. Especially in today's extremely tough economy, the pursuance of Lean becomes an easy target, inasmuch as it's often viewed as having nothing to do with effectively meeting production schedules, the principal task assigned manufacturing. But although that is true in one sense, it is absolutely false in another.

First

A good, aggressive Lean Manufacturing effort can result in saving the company money in the form of significant reductions in scrap, rework, obsolescence, work-in-process inventory, and setup and changeover, along with making vital improvements in direct labor efficiency and more. To some degree, halting or delaying the implementation of Lean to reduce expenses is like cutting off one's nose to spite one's face.

RELATED EXPERIENCE: I was working with a company, conducting a series of training and implementation events over an 18-month period. Some very substantial improvements were made and an ever-growing number of employees were exposed to the process. After I wrapped up the assignment, everything seemed to be on a positive track. Over the course of the time I spent in the factory I made some very good friends, including a number of the management staff, whom I frequently spoke with about progress and gave advice and counsel as needed. I came to find out afterwards that the company suddenly became faced with some very stern competition out of China. The feeling was that the competition was essentially striving to buy much of the market by fixing prices well below what it took to produce the products involved. In order to compete with the challenge, a decision was made to meet the competitor's price for a time.

In doing so, operating expenses were evaluated and a decision was made to cease any and all expenditures aimed at Lean Manufacturing. The Lean coordinator was taken out of his role to assume a shop floor supervisory position and from there all Kaizen and Lean-related activity essentially ceased. The Lean coordinator eventually moved on to another job, performing Lean training and implementation.

Unfortunately, I wasn't aware of the extent of the problem until far too many things were put in motion and didn't have the opportunity to plead the case with management, prior to the decision being made. But the company never regained the momentum it once had with Lean, although it managed to survive with a downsizing and focus on niche products for which the Chinese competition found little apparent interest in fighting.

Second

A decision to temporarily halt the insertion of Lean sends the wrong message to the workforce. Even if the process is actively resumed at some point down the road, it will usually be perceived as something less than fully important to management and thus vital to the success of the operation. Once a Lean initiative is underway, every effort should be made to keep it an active ongoing process. In fact, most operations would benefit, under tough economic conditions, to step up Kaizen-related activity.

It would of course be easy to conclude the experience noted was a unique circumstance, where the continuation of Kaizen-related activity and a focus on fully implementing Lean would have made little difference. On the other hand, as long as a company remains in business, downsized or otherwise, there is always the chance of it happening again, up to and including the possibility of being driven completely out of business. Therefore, before dropping or delaying anything that serves to assist a company in becoming more competitive overall, which Lean has been repeatedly proven to do, great care should be taken.

It's good to keep in mind that although Lean isn't something that always provides instant payback, it does provide the assurance an operation is on the right track. Implementing Lean fully and achieving the substantial benefits requires a long-term commitment and an unwavering effort, even under the most trying circumstances. In fact, in times of unusual or unique competitive pressures, the momentum of Lean should if anything be substantially increased.

The Do's and Don'ts Associated with Kaizen

As mentioned earlier, there are some definite dos and don'ts associated with a viable Kaizen process. Two of the more important don'ts were noted previously, but are included again in the following.

Definite Do's

1. Do communicate the full extent and scope of Kaizen, noting the intent is to make a full revision to the system of production and that the scope of the change will come to involve each and every employee (hourly and salaried) in the process.
2. Do appoint a full-time, highly qualified, and energetic Lean coordinator, along with an adequate supporting staff.
3. Do make the best use of the company's production engineering function, specifically with regard to enhancing key production equipment in support of Lean.
4. Do prepare an official Lean/Kaizen-related master plan, approved at the highest levels and used as a roadmap and timeframe for the full implementation of Lean practices and procedures factorywide.

5. Do make absolutely certain the maintenance function is properly staffed and fully capable of actively supporting a viable Lean Manufacturing effort.
6. Do construct formal written objectives in support of Lean for a select number of key players, aimed at creating their needed involvement and support. As training and awareness proceed, require an increasing number of employees to carry formal Lean objectives.
7. Do begin actively driving the use of Kaizen into the office arena and up the supplier chain.
8. Do formulate and introduce an incentive plan for Kaizen at an individual job level.

Definite Don'ts

1. Don't start a Kaizen event or activity unless maintenance is fully capable of supporting the effort.
2. Don't hold a Kaizen event or activity unless the right people are involved.
3. Don't leave the impression Kaizen is a discretionary activity that isn't absolutely vital to the success of the company.
4. Don't allow business conditions or other company undertakings to take precedence to the point of delaying, minimizing, or seriously disrupting the progress of Lean and associated Kaizen activity.

There are other dos and don'ts that aren't absolutely vital, but which serve to promote a more effective Kaizen process overall. These include:

Preferable Do's

1. Do, where possible, have the Lean coordinator report directly to the plant manager.
2. Do, where feasible, have the maintenance manager report to the Lean coordinator.
3. Do conduct at least one formal staff meeting monthly (chaired by the plant manager) that is exclusively aimed at reviewing Kaizen activity and the progress of fully inserting Lean Manufacturing.
4. Do incorporate regular management tours and audits of the process.

Preferable Don'ts

1. Don't turn down the opportunity to visit other factories using Lean and learn everything possible about what they are doing. No one has a monopoly on best practices and the process of learning should be an ongoing activity.
2. Don't become lax in making a conscious effort aimed at aggressively moving the process forward.

Simple Exercise for Getting the Most Out of Any Kaizen Effort

One of the most difficult things to do in a Kaizen event is to get everyone fully aligned with the task. Regardless of the type of Kaizen event being conducted, covering the "various versions of a process" on the front end of an event is one of the best ways of accomplishing the task (see Figure 3.3). Every process, regardless of whether it's in the office or on the production floor, has three versions. The first is what the process is thought to be, the second is what the process actually is, and the third is what the process should be.

RELATED EXPERIENCE: I once asked the production manager of a factory to outline on a flip chart precisely how a production process we were discussing worked. I then invited him to accompany me to the floor where we collectively went about accessing what was actually occurring. The differences we found were dramatic. We came to learn that operators had taken it upon themselves to change the prescribed method outlined by engineering. Some of the changes made were absolutely necessary, driven as we came to learn by a change in material by the purchasing function, which somehow failed to be related to both engineering and production management. But we also found that other changes made by the operators were in no way necessary and in fact served to increase the overall cycle time. To sum things up, it was found that the new material was less expensive but created problems with the equipment, which wasn't designed to readily accept it. The material was changed back to what was originally specified and work was done on correcting the method utilized by the operators involved, and the problems with the process disappeared.

Versions of a Process

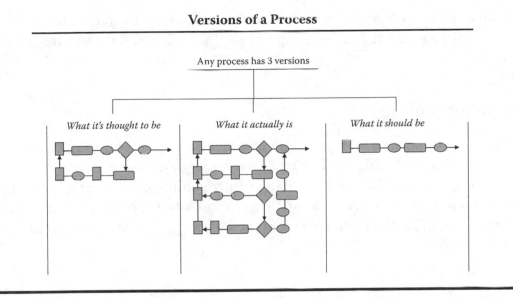

Figure 3.3   Versions of a process.

To make the point, allow the participants to choose a process and discuss and record what the process is thought to be. Follow this by taking the participants directly to the floor and recording what the process actually is. Sum up by allowing the participants to reach an agreement on what the process should be. By the end of the exercise the participants should generally be more knowledgeable about the task before them and much more open to how Kaizen can assist in achieving lasting improvement.

It's important to be mindful that most manufacturing facilities have much going on and communications and other facets of the business can sometimes fail to work as intended. As a result production processes on the shop floor can be influenced in a negative manner and become "problem processes." Thus there is all the more reason that the principles of Lean Manufacturing, driven by a viable Kaizen process, are fully incorporated and audited on a regular basis; and further enhanced with an effective strategy aimed at individual job improvement.

Key Summary Points
Staying Focused

Kaizen isn't something that can be left entirely to its own devices. It has to be coupled with a sound plan that is reviewed by senior management and followed up on a seriously active basis. It also has to be something that is focused on daily and reported on and reviewed just as frequently as to how well production itself is going or how budgets and forecasts are being maintained.

Avoiding Slippage

The very worst thing that can happen with a Kaizen initiative is to allow accomplishments made to deteriorate. Someone at the factory level has to be adamant about the need for change, fully committed to seeing it through, and unwilling to take "No" for an answer. Under the very best scenario that person would be the plant manager (see "Allowing Kaizen Accomplishments to Deteriorate").

Putting Lean Duties in Writing

It isn't practical to expect a firm that has long used batch production to replace its typical measurements and common objectives with those that are fully in tune with Lean Manufacturing. But changing or amending the written objectives of certain key players is vital to the overall success of a viable Kaizen effort.

Various Versions of a Process

Every process, regardless of whether it's in the office or on the production floor has three versions. The first is what the process is thought to be, the second is what it actually is, and the third is what the process should be (see Figure 3.3).

Removing Problems and Enhancing Individual Performance

Under typical manufacturing conditions production processes on the shop floor can gradually be influenced in a negative manner and become problem processes. Thus, all the more reason that the principles of Lean Manufacturing, driven by a viable Kaizen process are fully incorporated, audited on a regular basis, and enhanced with an effective strategy aimed at individual job improvement (see "Simple Exercise for Getting the Most Out of Any Kaizen Effort").

Vital Role of Production Engineering

There are three key areas where production engineering becomes a vital player in Kaizen. The first is in a qualified application of Poka-Yoke, where efforts are made to engineer improvements to equipment, tooling, and the like, which are aimed at avoiding common production errors. The second area is in the advanced application of SMED, where setup on key equipment in the factory is cut to single minutes, at a maximum. The third, but far from least important, is being the function employees can rely on for sound advice and direction in making improvements to their individual jobs (see "Failure to Effectively Utilize the Production Engineering Function").

Chapter 4

Where to Start and How to Proceed

Thinking Outside the Box

Robert Ballard, marine geologist and oceanographer, developed deep-sea surveying tactics that led to the discovery of the *Titanic*, the *Lusitania*, the German battleship *Bismarck*, and John F. Kennedy's *PT-109*, among others. He did it with the same basic technology being used by other deep-sea explorers but applied an entirely different strategy. Rather than using sonar to try to locate the vessel itself, he worked to find the debris field leading up to the vessel. It served to change the entire thinking about how to go about deep-sea exploration. When Ballard was asked, in an interview on CBS Television's *60 Minutes*, why he'd been able to achieve what others couldn't, his reply was, "Because everyone had trouble thinking outside the box."

Incorporating Lean in the most effective manner and working to make Kaizen a formidable competitive weapon starts with a willingness to move outside the box of conventional implementation practices. The best ideas and strategy will be of no avail if someone in a leadership position doesn't decide to take the opportunity to apply them. The important consideration is asking whether anything about it serves to restrict the plant's ability to adequately meet customer demand, in terms of cost, quality, or deliverability.

There is absolutely nothing about the outlined concept that can damage a manufacturing operation's ability in meeting its basic obligations to the business. In fact, all it can do is help. The single biggest hurdle is making a sound case for the need and then gaining approval as required for any added or extraordinary expense involved, which brings us to the topic of cost and payback.

Sum of the Added Cost and Payback of Lean

Something very few advocates of Lean like to discuss is the sum of the added cost a firm will have to absorb in order to make adequate progress with implementation. To a large degree the impression has been given that the cost of implementing Lean is basically insignificant. After all, firms are told, the work is aimed at improving existing equipment and facilities and by extending to the current workforce some hands-on training, a company can make the change required with very little expense. Unfortunately, this simply doesn't represent reality. The old saying that you can't get something for nothing applies to Lean as well as anything of real lasting value.

In a large manufacturing complex with a substantially large level of employment (500 or more employees) the expense for Lean, if done properly, can be as high as $250,000 annually for the first two years of the effort. The elements of cost potentially involve adding needed resources to the production engineering ranks, hiring a Lean coordinator and providing the coordinator with assistance (one to two employees), along with workforce training, and the expense of maintenance time and materials.

The added expenses noted do not take into consideration any offsetting factors, such as a company restructuring its typical annual training budget to direct the funds principally at Lean. Nor does it take into consideration reorganizing the salaried ranks to provide the new resources required without expanding overall salary headcount. If a genuine effort is made with respect to these, which is highly recommended, it potentially means some existing jobs would have to go in order to provide the resources needed for Lean implementation, without substantially expanding the operating budget.

This isn't always a comfortable step to take but there is sound reason and logic for the consideration of doing so, and it is why some solid planning has to take place before starting a Lean initiative. I've seen repeated cases where a start was made and the benefits of incorporating Lean became extremely apparent. But what also became apparent was the fact that in order to successfully spread the process across the factory it required manpower and expenses the plant wasn't equipped to handle. When this happens Lean often proceeds to die on the vine. Not an official death, of course; the process just slowly begins to crumble.

Although there is indeed an added cost associated with implementing Lean in an effective manner, the payback can be substantial in terms of greatly reducing wastes inherent to the system of production. In larger operations, the usual reduction made in work-in-process inventory alone will more than offset the added expenses noted. One can't forget if they carry a level of WIP inventory approaching $300,000 or more (in some cases, substantially more) that means a large ongoing investment is made just to open the doors of the factory and conduct business. Rid yourself of a large portion of that, which is what a good Lean initiative will accomplish, and the savings can be used for other more rewarding investment purposes, such as the cost of Lean itself.

Many of the benefits gained, however, are just as much related to "making money" as saving it. Under a good Lean initiative a plant should become noticeably more flexible and customer responsive. There should be a definite improvement in workmanship and product quality. Manufacturing lead-time should begin to show steady improvement; along with all the measuring sticks that lead to the potential of added business and additional profit.

As I've stated in every book I've written and won't pass up the opportunity to repeat again: No company can *save* its way to profitability. Think about that a moment. If an operation puts a pronounced focus on "saving" money, it's doing it for a reason. Most often that reason is due to panic because the existing way of doing business is running into problems. But no degree of reduction in the small things typically addressed in such an effort is going to make the difference needed. Where the premium focus needs to be is on what the company can do to make money, and that is where Lean can be of substantial benefit.

Savings opportunities alone under Lean, however, can be substantial. Assume there's a factory that has 150,000 square feet of manufacturing space. The plant employs 155 direct labor employees and 45 indirect. The cost of standard work-in-process inventory averages $300,000. The cost of scrap and rework runs $20,000 annually and downtime expenses equate to an average of $40,000. Using only these elements of the business for establishing a savings projection, the following apply as being highly potential under a soundly implemented Lean process:

■ Minimum improvement of 30% in space requirements = 45,000 sq. ft. of space opened for the procurement of additional business and/or new products, without the expense of brick and mortar. Using $60/sq. ft. as a conservative construction estimate = $2.7 million in potential cost avoidance.

■ Minimum improvement of 15% in productivity, measured in terms of the number of operators involved = $497,232 (calculated at a cost per year for wages and benefits of $19,968, using 24 operators, making $8 per hour + 30% fringe benefits).

■ Typical reduction of 80% in work-in-process inventory = $240,000, plus any associated inventory carrying charges.

■ Minimum reduction of 50% in scrap, rework, and downtime = $30,000.

Striving to be conservative with the savings noted—not including a potential cost avoidance of $2.7 million or assuming any reduction in indirect labor that normally happens—the savings for the improvements noted would equate to a sum of $737,262. Applied against a potential annual cost to implement of $100,000 a year over a three-year period, the ratio of savings to cost would be $2.46 for every dollar expended. That's a return on investment that most people would literally jump at making.

The factory example was purely fictional, however, the cost savings potential is far from that. The reduction percentages outlined have been proven time and again in factories around the world. In fact they represent a relatively conservative estimate of the actual improvement potential. Anyway one chooses to look at it, the money spent on fully implementing Lean will always show a solid payback on investment—if a proper application of Lean is applied.

Progressive Kaizen Tool Box

Progressive Kaizen, as pointed out in Figure 1.3 (Chapter 1) is one of the three major components of ALIP (Advanced Lean Implementation Process). Figure 4.1 provides an overview of the tool box utilized for progressive Kaizen, which starts with four guiding principles: workplace organization, uninterrupted flow, error-free processing, and insignificant changeover. Outlined for each drawer of the *Component Tool Box* shown are the specific processes or steps involved in achieving an ultimate level of application.

Outside of a conscious focus on the four guiding principles, the other critical components are discretionary management initiatives, the effective use of production engineering, business process improvement (which is

Progressive Kaizen
Component Tool Box

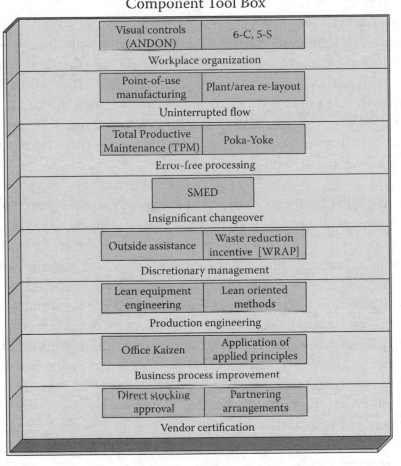

Figure 4.1   Progressive Kaizen component tool box.

aimed at moving at some point into the office arena with the intention of significantly improving key business processes and driving the use and benefits of Kaizen down to the individual job level), and finally placing a focus on vendor certification standards that serve to support and enhance the process. Each component has been or will be spoken to, but Figure 4.1 can be a helpful reference.

In examining the specific tools involved in more detail, the following are noted:

- *Workplace Organization:* Applying the guiding principle of workplace organization to Kaizen-related activity involves utilizing 6-C or 5-S (6-C being a somewhat more all encompassing version of Toyota's widely used 5-S process), along with an extensive application of visual aids and controls.

■ *Uninterrupted Flow:* Applying this guiding principle involves a focus on point-of-use manufacturing techniques, along with laying out the area involved anew to bring operations closer together and eliminate as many stop and storage points in flow as possible. This would include utilizing the concept of U-cell arrangement, where feasible. At some point the factory should also be completely rearranged to support continuous flow and take advantage of the space reductions gained through ongoing Kaizen activity.

I often use what I refer to as the "Swiss cheese phenomenon" (see Figure 4.2) to explain how to take premium advantage of space gains achieved through good Kaizen activity. As the process goes forward, areas of space will be opened, some of them often small and insignificant on their own. But any and all space opened as a result of Kaizen should be lined off and considered to be unusable by anyone, until the accumulation of the added space can be used to layout the entire factory anew and open a section of the plant for new products and equipment, along with other potential business opportunities. I've always told operations that space freed up as a result of Kaizen should become the property of the plant manager and that no one should have the right to use it for any purpose until a total factory rearrangement is made, unless of course the plant manager gives express permission to do so for an appropriate short-term purpose. This does a couple of important things:

1. It provides a clear visual example of the importance of Lean: opened space that employees walk by each and every day and come to understand has a distinct long-term purpose in advancing the business.
2. It shows that the leader of the factory takes a strong personal interest in Lean and shares a commitment to making it a lasting reality.

■ *Error-Free Processing:* Applying this guiding principle involves the effective use of Poka-Yoke and the supporting advantages of TPM on all equipment and production processing throughout the factory.

■ *Insignificant Changeover:* The use of this guiding principle centers on reducing setup and changeover to the point of setup becoming "insignificant" to the decision-making process, for such things as taking on

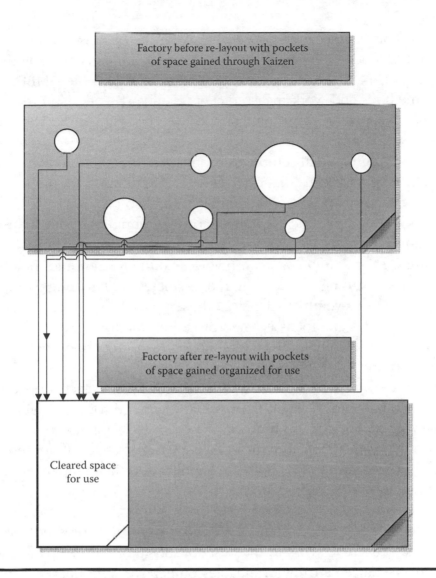

Figure 4.2 Factory conversion: "The Swiss cheese phenomenon."

added business, new products, and so on, through an expanded and professional application of SMED.

■ *Discretionary Management Initiatives:* These involve supportive management decisions that enhance and promote the implementation of Lean and make the tasks involved with Kaizen less complicated to implement. It involves such things as incorporating a WRAP incentive (or bonus) for individual job improvements and calling in a qualified consultant for special training or assistance, along with special project-based improvements such as investing capital in new or revised equipment in keeping with good Lean practices.

- *Production Engineering:* Concerns making the best use of the company's production engineering function, which means utilizing their expertise in a highly advanced application of Poka-Yoke, SMED, TPM, and methods and work measurement ("standard work") in order to ensure equipment meets good supportive Lean standards. The standards would include:
  - Achieving minimal setup and changeover on all equipment
  - Building in devices that catch defects due to material, the incorrect orientation of parts and components, and so on
  - The creation of efficient, lowest-cost operating procedures through sound methods analysis and work measurement practices
- *Business Process Improvement:* Involves taking Kaizen into the office arena and focusing on reducing redundancy, standardizing work, and improving the overall efficiency of those performing office duties. In doing this, a prime focus is again placed on the four guiding principles that apply as readily to business functions as production operations on the shop floor.
- *Vendor Certification:* Concerns officially certifying select suppliers to deliver directly to the shop floor, without going through a company's receiving and inspection function. The certified supplier comes to have direct stocking approval, which in most cases involves a partnering arrangement favorable to both the supplier and the company, in terms of pricing, volume, and delivery.

All of these elements of progressive Kaizen must be addressed and utilized for the process to work to the best advantage of the company and serve to make Kaizen a truly formidable competitive weapon.

Advantages of Labeling Kaizen Activity "Waste Reduction"

The word *Kaizen* is a foreign term in the United States and is often less than easily understood by the average American worker. Even after training some are not sure how to fully align themselves with the task prescribed.

Tell a group of average American workers you want them to perform Kaizen and assurance is granted that many will not relate to the task, even after some exposure to the process. But tell them instead you want to engage them in an activity associated with waste reduction and they will immediately relate to the mission. They may not fully understand the particular wastes the exercise is aimed at eliminating until they've been given proper training, but they will clearly understand the meaning of the task involved.

It is helpful to remain mindful that Kaizen is the *tool* used as an accepted process to make change and perform continuous improvement activities. However, in the mind of the average American worker the word doesn't easily translate to the actual *task* without a good deal of explanation. That task, of course, is waste reduction.

In most Lean initiatives an effort is made to create the understanding that Kaizen means continuous improvement. Although that is true in the sense of a broad translation, it simply doesn't carry the impact for most workers that *waste reduction activity* does, nor does the word itself serve to inspire a conscious focus on the job that all employees should hold in assisting the company to become more competitive. However, couple the words *waste* and *reduction* with *activity* and *process* and you have "Waste Reduction Activity Process" (or WRAP) which serves to adequately describe what the effort is actually aimed at accomplishing: "wrapping" the implementation of Lean in an active and ongoing waste reduction effort, aimed at making sound continuous improvement.

RELATED EXPERIENCE: An interpreter in Germany proceeded to point out to me during a Kaizen event I was conducting that I was confusing everyone with all the Japanese terms and went on to say, "In speaking for the interpreters, if I were you, I'd stick to English so the words can be translated in a manner everyone can easily understand. If 'Poka-Yoke' means mistake proofing, just say mistake proofing and leave it at that." It was obvious the interpreters were principally concerned with having to translate a somewhat confusing mix of Japanese, English, and German for a subject with which they were totally unfamiliar. On the other hand, the thought remained with me as I went about writing my first book some years later, *Fast Track to Waste-Free Manufacturing*. In that work I made a conscious effort to speak to the four guiding principles involved in English terms; for example, using "Error-Free Processing" rather than identifying the entire process of mistake proofing as Poka-Yoke. The same held true for "Insignificant

Changeover," which utilizes SMED as a primary tool, along with "Workplace Organization" (through the use of 5-S, 6-C, and Andon) and "Uninterrupted Flow" (utilizing One Piece Flow and Point-of-Use Manufacturing techniques.)

Later, as I went into consulting work, I experimented with the importance of taking this step by asking participants at various Kaizen events to help perform an exercise. I began by writing down the words Poka-Yoke, SMED, 5-S, and Andon on a flip chart and then proceeded to ask what those words meant to the audience. The response was a mix of considerable confusion, even from those who'd had some previous exposure to the terms. I then wrote the words "Insignificant Changeover" along with the other three principles of waste-free manufacturing on a flip chart and asked for the same response. It was dramatically evident that the audience much more quickly associated with the terms and came to more easily understand their relationship to the job at hand. I made the practice afterwards to always identify the *tasks* involved in the terms of the principles noted and then moved from there to explain which *tools* of the Toyota Production System were used to make the appropriate change. On more than one occasion, the participants commented it made the overall process much easier to relate to and to understand.

The job of making a broad sweeping change in the way a factory goes about performing the mechanics of production is tough enough in itself. Therefore, keeping things in terms the average worker can easily comprehend could be much more important than it might first appear. It's far easier to start with a process defined in English and reference where the appropriate tools of TPS apply. No less formal credit is given to the particular tool involved with this fashion of definition than the other way around.

While firms that have started a Lean initiative might be reluctant to extend an effort at changing any reference of Kaizen to waste reduction; this is an area where they shouldn't apply a deaf ear. Adopting the practice of using waste reduction as a watchword can be done without making a formal announcement of an intention to do so. Simply begin using the term as frequently as possible and gradually allow it to become what everyone eventually perceives the process to be.

Tell an employee his help is appreciated in advancing Kaizen and in all probability he'll see it as nothing all that significant. On the other hand, most employees take a great deal of pride in being recognized for eliminating wasteful practices. This is because most have been taught from childhood to avoid being wasteful. With respect to the overall job to be done this could be considered a small thing, but in the scheme of making accomplishments in the fastest and smoothest manner possible, centering the focus on "waste reduction activity" can make a noted difference in the mind-set and active support of employees.

Value of Putting the First "Pull Zone" in Final Assembly

I was once asked what single thing a company could do that would serve to get everyone's attention regarding the task at hand. My response was make the entire final assembly portion of the factory a mandated "pull zone" and ensure there's strict compliance.

A plant's first mandated pull zone should carry two distinct purposes: the first is to train the workforce and provide hands-on experience in the concept: the second purpose is to do something that strongly encourages (or essentially drives) the advancement of Lean.

In many cases the decision regarding where to start a Lean initiative on the shop floor is based on where it's felt the least amount of disruption to normal production will occur. This is understandable to a degree. However, an important factor that should be considered, above everything else, is where a viable "pull zone" cannot only be implemented, but will stand as the leading example of where the factory is headed in the future. With this in mind, one or more production lines in a plant's final assembly area become the ideal place to start. There is an important reason why.

Final assembly is the spot where the greatest amount of the work performed in a factory is funneled. As a result this area typically becomes a gathering place for most of the parts and components produced in a factory and therefore an ideal spot to ensure that unless it is "called for," nothing is received. Doing this builds the understanding that no matter how various production areas in the factory may be producing what they deliver, they are only allowed to send it to final assembly in the quantities specified by the user. Until it is eventually called for, the remaining inventory becomes the producer's job to manage.

Very few factories take such a step, largely to avoid having to deal with precisely what to do with the excess inventory involved. Where would it be stored? Who would be responsible for overseeing it? Where would the added space required be obtained? But imagine, if you will, the influence this could have on directing everyone's attention to Lean and in helping to change the production mind-set of "produce and push" inventory down the line. Under the "push" scenario, once inventory is out of sight, it essentially becomes someone else's problem to deal with and out of mind.

After a plant's first pull zone has been established for a given final assembly line, the next objective should be to see the approach is fully incorporated throughout final assembly, as a whole. Implementing a strictly adhered-to pull zone essentially forces the supply chain to work at reducing excessive inventory, where it can.

This one action alone raises the attention level on Lean and often motivates supervisors to personally approach the Lean coordinator and ask for assistance in how they can lower the inventory level they're carrying. But regardless of whether this indeed comes about, in most cases it would not be long before supervisors of a plant's feeder areas would be much more open-minded and enthusiastic about getting the principles of Lean Manufacturing applied to their particular area of responsibility.

This type of change obviously should not be incorporated before some appropriate communications are conducted; which could go (in part) something similar to the following.

> We're here to discuss the need to change our general approach and overall technique to production. Most of you have heard and seen, through various meetings and written communications, that we are pursuing a Lean Manufacturing initiative for the factory. This obviously won't happen overnight and will require the support of each and every employee. But in order to take this important step, one of our four major assembly lines has been selected as the plant's first "pull" zone. What this means is that parts and components can only be sent to this particular line in the quantities specified by the user. The remaining inventory will have to be stored in the areas where it's produced until it's eventually called for. For a time, this will undoubtedly place an added level of work and responsibility on various areas of the factory. But it is important we do this and we won't back away from doing everything necessary to make it a full success. As time goes by we will extend the approach to other assembly lines and work to implement the kind of change that will make it easier for every production area to meet this requirement.

Going about this properly would not only meet the first objective noted for the plant's first pull zone, but also the second, which is to strongly encourage and essentially drive needed change through the entire factory. Unfortunately, I have seen repeated cases where a Lean coordinator stood ready to perform Kaizen and aggressively work at advancing Lean, but found it extremely difficult to find managers and supervisors who were keenly interested in pursuing Lean in their areas of responsibility. In fact, in most cases the complete reverse was true.

Up until most manufacturing managers and production supervisors have a pressing need to do so, they will generally strive to avoid the challenge of a formal Kaizen event. This is understandable considering they are still being measured from the standpoint of how well they comply with the rules of the existing system of production. But given they are essentially forced to deal with a mandate for nothing to be sent to a specified area or department unless it is specifically called for, attitudes will start to change dramatically.

Making final assembly the first fully mandated pull zone for the factory will undoubtedly create some confusion and frustration, but in the end it can serve to place a growing level of attention on fully implementing Lean Manufacturing as fast as possible throughout the entire facility.

RELATED EXPERIENCE: I once proposed to a plant manager I was consulting with that he should have a pull zone concept established for the entire final assembly area; which consisted of three large, highly paced assembly lines. His reaction was how could he possibly go about doing that without laying the appropriate groundwork in the backup areas, in other words, without first applying "Lean" to all the feeder areas of the factory. My response was, "That's easy. Just do it." I went on to explain that although it could initially create some confusion and possibly some frustrations; it was the best thing he could do in order to get and keep everyone's attention on the task facing the operation; and went further to say I would give him any help and assistance he needed in making it happen.

He simply couldn't bring himself to include the entire final assembly area, but agreed to try it on one of the lines. He proceeded to communicate the need to those involved and began the practice. True to form, there were questions as to what to do with the added inventory various areas were required to deal with and hold.

He was struggling with a solution when I pointed out to him that the plant had 12-foot (wide) aisles throughout the entire facility; and although this gave a nice appearance, 8-foot aisles were perfectly acceptable for conveyance purposes. A simple 4-foot stripe was added inside the outer marker of the existing aisles and became the place where excess inventory was stored until it was needed. Many area supervisors didn't like the arrangement for appearance purposes, but any excuse regarding the lack of a place to store the excess inventory involved was eliminated. The plant manager went on to emphasize to anyone who chose to complain, "I don't like it either, but it's something we have to do for the time being. My advice is for you to get with Joe (the assigned Lean coordinator) and be one of the first to do something about it."

The plant manager turned out to be one of the best manufacturing leaders I've ever worked with and it probably goes without being said that the factory went on to make some outstanding accomplishments.

Sticking to the Plan and Avoiding Disruptions

Any plan is only as good as the will to see it through. There will be distractions with any Kaizen process that draws resources away from advancing Lean and implementing the kind of change needed. I have seen repeated instances where an extensive amount of work was done in a factory in order to establish a showcase area designed to represent where the factory intended to go with Lean Manufacturing and little was done afterwards to actually make that the case. There are many reasons why disruptions to the implementation of Lean occur. But a few of the more common distractions involve:

1. A failure to fully understand the importance of seeing the change as being absolutely vital to the overall success of the operation
2. Other priorities and initiatives that direct attention away from the task
3. Changes in general management or the reporting and organizational structure

As it happens, the last distraction is one of the more prevalent reasons a Lean initiative sometimes stalls or in some cases is completely abandoned.

During my career in manufacturing, something that became extremely noticeable was the tendency for change in plant leadership. It seems to happen every two to three years on average. If it didn't occur at a plant manager level, it occurred in other senior leadership positions.

New leadership always brings new thoughts and most often new direction, among other influences that can sometimes inadvertently or otherwise stall a Lean Manufacturing effort. The best way to guard against this is to have a formal documented strategy for Lean, including a master plan for Kaizen, along with a structure for the process that promotes appropriate audit and follow-up procedures. Doing this avoids having to rely on someone to serve as the individual guardian of change. The latter method can work given the individual in charge of the effort is someone who stays around long enough to see it through. But what is better is a process that serves to guard against stalls in implementation, through the power and influence of a well-established structure for change.

Having this in place aids in selling the importance to new management by indicating to them there's a well-structured plan established, along with evidence that progress is being made. Unfortunately, no amount of proof will make a difference if new management doesn't hold an appreciation for Lean or at least a willingness to hear someone out and make a sound judgment based on the information provided. Regardless, the more a manufacturing facility can make Lean a standard way of conducting business, just as with meeting production schedules, achieving budgeted forecasts, and the like, the better overall chance there is of it carrying the influence needed to survive disruptions that can distract from the ultimate goal.

Conducting the Factory's First High-Impact Kaizen Event

A factory's first High-Impact Kaizen event will in most cases set the stage for precisely where an operation is headed in making change for the better. It therefore requires a good deal of forethought as to precisely what area of the factory will be the recipient of such change and who the participants for the event will be.

The purpose of a High-Impact event is to make sweeping change to an area of the factory that incorporates as many aspects of Lean Manufacturing principles as possible during the exercise. The length of such an event can and will vary, but would normally be one to two weeks in duration, with approximately one half of the time spent on intensive training in Lean. The number of participants would range from 25 to 30, most of whom should be mid- to high-level managers, along with a select group of hourly and salaried contributors. Those who should definitely be considered as full-time participants are

1. The production manager
2. Two to three key production supervisors, including the supervisor over the selected area
3. At least one representative from the company's maintenance function
4. Three to four key hourly production associates
5. The production engineering manager and a majority of the production engineering staff
6. Two to three salaried employees from functions normally detached from production work
7. At least one representative from the labor union, should one exist

Basic Event Objectives

There are a number of goals that should be explained and established for the event on the front end. These include, at a minimum:

■ Improvement in floor space required of 20 to 30%
■ Reduction in standard work-in-process inventory levels of 70 to 90%
■ Improvement in productivity (measured in total number of operators) of 10 to 15%
■ Improvement in measured quality indices of 30 to 50%

These will normally be perceived by the group involved as being next to impossible to achieve, but they will quickly come to see that it indeed can. A high level of focus should be placed on these objectives and as change proceeds, the group should be asked to report on how things are progressing against the assigned goals of the event.

Participation of the Production Manager

Under no circumstances should a plant's first high-impact Kaizen event be conducted without the production manager serving as a full-time participant. This doesn't mean he has to take a temporary leave of absence from normal activities during the course of the event, but it does mean he has to play a very active role and be personally involved in the overall change formulated and conducted. The production manager's participation has to be extended without being dictatorial about what change will or will not occur. A word of advice that should be given production managers is for them to allow the process to work as intended, which will require them to avoid pushing preconceived notions on the participating team.

RELATED EXPERIENCE: In a high-impact event I was conducting for a manufacturing firm, it quickly became apparent that the production manager was intent on placing more emphasis on protecting the status quo than making change for the better. I approached the plant manager and passed on my observation, noting there were two ways we could approach the matter. The first was for him to speak with the individual and in his own way convince him of the need for an open mind and his personal encouragement of the thoughts and ideas of others. The other approach was to allow me to speak with the production manager and see if I could somehow sway his thinking, without it having to go as far as the plant manager intervening. We collectively decided on the latter and I proceeded to pull the production manager aside for a friendly cup of coffee. Once the conversation was underway it went like this.

"Fred, you have a very forceful personality and that's a plus for the job you hold," I began. "But if we're not careful, it can also be a hindrance to what we're trying to achieve."

"What are you driving at?" he asked, appearing to be bit perplexed.

"You have a great deal of knowledge related to what's going on in the factory. But along with that comes a tendency to quickly deflate the ideas of those who have less overall experience. If this continues, it won't be long before things get to the point where the only idea that matters is what you think is best."

He smiled and remarked, "So what's wrong with that?"

"I'm going to be perfectly frank and I'd ask that you don't take what I'm going to say the wrong way," I replied. "My job is to strive to get the best out of the event and demonstrate the considerable advantages of Lean. Whether you believe that's the right direction to go or not, it's good to remember you didn't personally choose the method of production you're currently in charge of managing. You simply came to work under its influence and you certainly carry no obligation to defend it."

He leaned back and thought for a moment before responding. "So you're asking me to just sit back and allow them to pursue things that aren't going to work."

"Not at all," I responded. "I'm simply asking you to place some reluctance on quickly deflating the ideas of others. That's it. I can assure you the process isn't going to allow something that doesn't work to the full advantage of the company. But you'll have to trust me on that. Nothing is going to be done without your full awareness and input. But we need a free flow of ideas in keeping with the principles outlined if we're going to make the event a success. Just give it a chance to work and see what happens."

Although it was difficult for him, the production manager went on to make a sincere effort to avoid downplaying the ideas of others. He would slip in the effort occasionally, but one could see him catch himself and back away before it became a hindrance. In the end he came to agree with most of the changes the group outlined and provided the support needed to see that they were implemented as fully as possible. Although I can't say he became a strong proponent of Lean, I witnessed enough to say he wisely decided against being viewed as a roadblock to the effort.

For those who desire a full and effective transition to Lean, the production manager would ideally be—or come to be—a strong proponent of Lean. Unfortunately many production managers come from the old school. They grew up under a batch manufacturing environment and some find it difficult to support any serious challenge to that technique. But most of them can and will adapt if management makes it absolutely clear the position holds a specific obligation to making Lean a full success and that nothing short of that is deemed acceptable.

Participation of Shop Floor Supervisors

A high-impact Kaizen event is an opportunity to take two to three key shop floor supervisors and provide them with extensive training in the principles and techniques of Lean Manufacturing. One of the supervisors who should definitely be selected is the person in charge of the area where change is going to be made. But spreading the depth of knowledge gained in a high-impact event to other areas of production leadership is just as important. Due to the expense and the time involved, high-impact events are not something conducted on a highly frequent basis. It is therefore to the distinct advantage of a company to spread as much of the experience as possible to a company's shop floor supervisors.

Participation of a Key Maintenance Representative

It is extremely important to have at least one key representative from the maintenance function as a full-time participant in the company's first high-impact Kaizen event. Ideally this would be the maintenance manager or supervisor, although it isn't always practical for this to happen. The representative, however, serves as the team's direct interface with maintenance and in aligning when and how changes proposed by the team are carried out to full completion during the scope of the event.

With regard to the issue of the full completion of work during the event itself, it's important to mention what the basic goals and measurements of success should be:

■ It should be made clear to the group on the front end that the only thing it can take credit for is what is fully implemented over the course of the event itself. Anything left undone or incomplete is credited to those who follow up later and make change accordingly.

■ Establishing a sense of urgency is of paramount importance to any Kaizen event, but that is especially so in a high-impact event. It's a time when all the appropriate decision makers are brought together as a team and when there is little excuse for not moving forward in a highly aggressive fashion. But any change made doesn't have to be perfect. Perfecting change, as closely as possible, should be the goal of sustaining Kaizen activity. As an example, most high-impact Kaizen events focus on establishing numerous visual aids and controls. To begin these can be something as simple as a piece of whiteboard with words or instructions printed by hand. These can always be enhanced through further sustaining Kaizen activity, for as a Japanese group of consultants once highlighted to everyone, "Crude and quick is better than slow and fancy."

In reality, the only thing that counts is what the group is able to achieve during the event itself, which avoids its turning into a lengthy exercise on a wish-list of items for the future, which serves to accomplish little.

What complicates matters somewhat is the fact that no one knows for certain what the actual changes will be until the first week of the event is well underway. In addition there is often the need to make corrections to various changes made, as the event proceeds. It is good to remember that the event is a learning process and corrections that serve to further enhance a production process are sometimes necessary. This leaves the need for a maintenance function that is highly flexible and extremely supportive during the event.

Participation of Key Production Associates

In order to make a high-impact Kaizen event a full success, three to four key production associates from the area selected to be revised are crucial. But there are a number of factors that should be taken into consideration regarding the participation of production workers:

- The changes made should in no way be viewed as something forced on production workers without an appropriate level of input from those who end up inheriting the outcome. Some of the very best change happens when the production workers involved with living with the change clearly feel it was of their own making. In addition, it is usually something they typically take a great deal of pride in accomplishing.
- An effort should be made to select production employees who have a broad level of experience, such as those fitting a "lead-man" or "relief" classification, who carry expertise in almost every job in the department. Far too often companies go in the opposite direction, selecting employees who carry little overall expertise and just as important the respect of fellow workers.

Participation of the Production Engineering Manager and Staff

Most manufacturing operations of any size have a production engineering staff. Smaller operations tend to have one or more production engineers reporting to someone, such as the production manager. In the case where a production engineering manager position exists, the assigned individual should be a full-time participant in a company's first high-impact Kaizen event, along with as many of the production engineering staff as possible. Under the best circumstances, a large majority of the production engineering staff would participate in the company's first high-impact event, with the exception of what is needed to keep the factory operating normally. A good goal is to strive to have at least 50% of the PE staff free to attend, which can be done in most cases with proper planning.

High-impact events can be conducted while at the same time maintaining the support needed to meet established production schedules if certain aspects of the event are properly addressed on the front end and communicated accordingly. A couple of pointers regarding this follow:

■ During the first week of a high-impact Kaizen event (normally two weeks in duration) time is spent training participants and readying them to make significant change on the shop floor over the course of the upcoming weekend. The start time for the first week should be one hour after the shift begins, with wrap-up one hour prior to shift end. This provides six hours of training each day for the participants, along with two hours (one hour on both the front and back end of the day) to address normal business activities.

■ During the second week the participating team is turned loose to make the planned changes and should be expected to spend full time in the course of accomplishing the outlined change. The participants therefore essentially have to be excused from sustaining responsibilities for the entire second week, unless an absolute emergency arises. Even if an emergency does arise, great care should be taken in pulling participants out of the event. But if deemed absolutely critical to do so, every effort should be made to see that they return to the team as quickly as possible.

Participation of Salaried Employees Detached from Production

The long-term goal again is to train the entire workforce, which includes salaried employees who are detached from the day-to-day business of running production. This applies to clerical and support personnel assigned to various office functions. Planning on having at least two or three employees from functions such as accounting, human resources, and the like at the plant's first high-impact event is important in spreading knowledge and encouraging future participation in the process in the office functions. In addition, such individuals often bring many fresh ideas to the table that otherwise would not be given consideration.

Participation of Local Union Officials

Should a labor union exist, every effort should be made to obtain their participation at some level during a high-impact event. Ideally, one or more of the union officials or a direct representative would participate. However, if for any number of reasons the union is reluctant to do so, they should be invited to attend the daily wrap-up sessions, where the change and pending results are discussed. Most union officials would agree to this and, in fact, would likely be pleasantly surprised they were asked. There is absolutely nothing in a Kaizen event that should be objectionable from a contractual standpoint, inasmuch as the change involved is structured to make a shop floor employee's job far easier and far less complicated.

If an event is run properly, the union will normally come away in support of the changes made. This in turn establishes the opportunity to enhance union/management relations and gain the support of the union in making Lean a full reality. In the case where an extremely hardened relationship exists between management and the union, a high-impact event provides the opportunity to make a positive step behind which both parties can rally. The biggest issue to overcome is providing assurance that operators who may be displaced as a result of the event, will not be laid off, which calls for a special series of comments:

■ *Defining "Displaced Operators":* What "displaced" means in this particular situation is discovering through the use of Kaizen that a particular job can be done with fewer operators. Those removed from the production process as a result are considered *displaced*. There are two specific issues that companies need to keep in mind.

- *Issue One:* The first issue has to do with assuring the workforce that for a prescribed period of time (six months to a year) no one will be laid off as a result of Kaizen and the overall implementation of Lean. If a downturn in business happens to occur, there has to be assurance the company stands firm behind its promise. In doing so, there has to be a means established that is capable of clearly showing that any layoff of personnel is strictly volume related, which in most cases can be readily done.

- *Issue Two:* The second and more sensitive issue deals with the long-term impact on displaced operators. No company can afford to carry personnel for an indefinite period that's in excess of what they can effectively utilize. There comes a time when it's necessary to adjust manpower in accordance. Most unions and hourly labor employees understand this. And although it is not the most pleasant topic to discuss, they are generally willing to sit down and seriously address what is fair and reasonable.

■ *Making Kaizen-Related Workforce Adjustments:* Once a company has clearly shown its intention of maintaining an acceptable labor pool made up of displaced operators, a layoff pertaining to the size of that pool may at some point be required. The company should make it clear on the front end that once a year it will perform an evaluation and make necessary adjustments in manpower pertaining to the established Kaizen labor pool. This doesn't negate the fact that some employees may end up being laid off as a direct result of a Lean initiative, but it clearly indicates that manpower reductions are not the principal purpose behind the effort. In most cases if a Lean initiative is structured and run properly, the chance of added business circumventing such an adjustment is reasonably good, although it isn't something that can be absolutely guaranteed. An additional factor that can aid in reducing the overall impact, however, is common attrition, which would allow operators in the Kaizen labor pool facing potential layoff to be absorbed in the standard labor pool.

■ *Being Fair, Just, and Forthright:* The answer to the sensitivity of this issue is for the company to be forthright about its intentions and fair in dealing with displaced operators. However, should a company hold workforce reduction as one of its principal goals, all bets are off the table. Management taking this stance should remember what they are effectively saying to the workforce is: "We want you to assist us in implementing Lean so we can reduce manpower and improve profits." To that I could only ask how those delivering such a message would feel if they were given the same option. It's safe to say less than confident about the pending success of the venture. Lean Manufacturing, if done right, requires a unique partnering of management and labor in a special effort focused on making a company more competitive. There is no room for shortsighted cost reduction and anything less than a show of appreciation for the efforts employees extend toward advancing Lean. The cruel fact is there usually comes a time when adjustments in manpower are absolutely essential. But it should be done in keeping with something other than simply making a short-term cost reduction.

- *A Common Flaw in the Use of Displaced Operators:* Something very noticeable when it comes to displaced operators is the lack of effectively utilizing this important resource. If an agreement has been made to hold such operators for a prescribed period of time, they're most often thrown into a general labor pool and used for relief and absenteeism purposes. A much better use would be to give this particular resource to the Lean coordinator, who could use them to assist in making further improvements throughout the factory or in relieving others to attend a working Kaizen session.

- *Creation of Two Separate Labor Pools:* A company's general labor pool should stand on its own merits and ideally would not include operators displaced as a result of Kaizen. Companies need to be more creative when it comes to Lean and a proper application of assigned resources. The biggest need is to learn to separate standard layoff issues from those that arise from activity aimed at making the company more competitive. In keeping with this, displaced operators should be viewed as an added resource to further advance that objective. Lean Manufacturing calls for change and the change encompasses more than just the production arena. It involves willingness on the part of management to revise policy and procedure as required, in order to make Lean a full and lasting reality.

As a visual representation of how the two separate labor pools should work, Figure 4.3 provides an overview. On occasion serious needs could arise in adequately meeting customer-related schedules; and at the request of the company's production manager a decision could be made by the Lean coordinator to allow the use of the Kaizen labor pool on a temporary basis. However, care should taken to ensure this doesn't become the rule, rather than the exception. This serves to point to the fact that the production manager and the Lean coordinator need to have a good working relationship and remain cognizant of the job that each is striving to achieve.

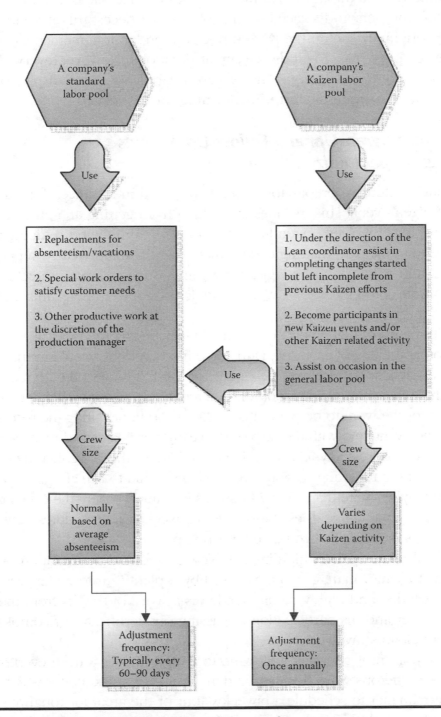

Figure 4.3   Inclusion of separate Kaizen labor pool.

It should be noted that in making the most effective use of a Kaizen labor pool, the person displaced from a job doesn't necessarily have to be the body originally sent to the Kaizen pool. Union labor agreements could stipulate seniority rules in all displacement situations. However, every effort should be made to have people in the Kaizen labor pool who display both the ability and a strong interest in performing Kaizen-related work.

Preparatory, Wrap-Up, and Follow-Up Aspects of a High-Impact Kaizen Event

Some rather extensive preparatory work is required in advance of a high-impact Kaizen event. This includes everything from getting an agreement on the participants involved, extending invitations, obtaining and setting up an adequate training area, making certain that essential maintenance support is available, and conducting the logistics for things such as breaks, lunch, start and stop times, and the like. A word of advice is not to underestimate the extent of the planning involved.

Something every high-impact event should include is a reporting and wrap-up session at the end of each day. During the first week this would include all participants and would normally be conducted the last hour of the normal work day. In this session the teams report on progress made in establishing goals and objectives for the work to be performed on the shop floor during the second week of the event, and with the help of the Lean coordinator and management in attendance, address any related problems and issues that arise. During the second week the reporting session is normally held mid to late afternoon and is approximately no more than 15 to 20 minutes in duration. Here the teams report against a set of written goals established during week one and address any serious problems or issues that could be hindering progress.

The final wrap-up for a high-impact event is a formal presentation by the teams to senior management, followed by a plant tour to review results. Usually included and highly recommended is a word of thanks from the plant manager and his participation in passing out certificates of completion to the participants involved.

Very important is adequate follow-up to a high-impact Kaizen event. There are numerous ways this can be done, but one of the better is for the plant manager to schedule a biweekly tour of the area, for roughly a two-month period, that includes all of his or her direct reports, along with the Lean coordinator. During the tour, the group looks to ensure slippage hasn't occurred and makes note where further opportunities exist. The tour should also include a walk-through of the general factory and some discussion as to where some immediate improvements, in line with good Lean Manufacturing practices, come to mind. The whole idea is to ensure Lean is given appropriate attention recognition and clearly says to the workforce that management truly cares about Lean.

Basic Structure and Steps Involved in Conducting a High-Impact Kaizen Event

A constructive high-impact Kaizen event should have a defined structure to follow in order to get the most out of the effort and provide the kind of overall training needed. The idea again is to touch on as many key aspects of Lean as possible in constructing an area in the factory that will serve to represent where the plant is headed in the future. Figure 4.4 outlines the process involved in conducting a high-impact event, noting where and when the four guiding principles apply and the sequential steps involved in making change on the shop floor.

The first guiding principle on which the team should focus is workplace organization, which is the one that should be carried out to the fullest during the course of the event. The tools involved are 6-C (or 5-S) and a strong application of Andon (visual controls.) Before the team starts making change, however, a "disposition zone" is established, where items that appear questionable (excess inventory; small, easy-to-move equipment; hand tools; storage cabinets; tables; chairs; and such) are physically removed from the area and placed for final disposition, which occurs just prior to the conclusion of the event. Letting the operators involved know that anything removed is in a safe place and can be retrieved if needed avoids creating concern and anxiety about things that may initially be considered absolutely required, but before the event is concluded will likely be understood as doing little but taking up valuable space.

Ideally, the participating group is broken down into teams that focus efforts on one of the various guiding principles. Workplace organization and uninterrupted flow will carry the largest number of participants; but certain select individuals (usually the participating production engineers) are assigned to address, recommend, gain consensus, and implement a sound application of SMED, Poka-Yoke, and TPM, on at least one piece of critical equipment in the area. Therefore, some of the steps noted in Figure 4.4 will run in conjunction with one another. But if any question exists regarding where work should be

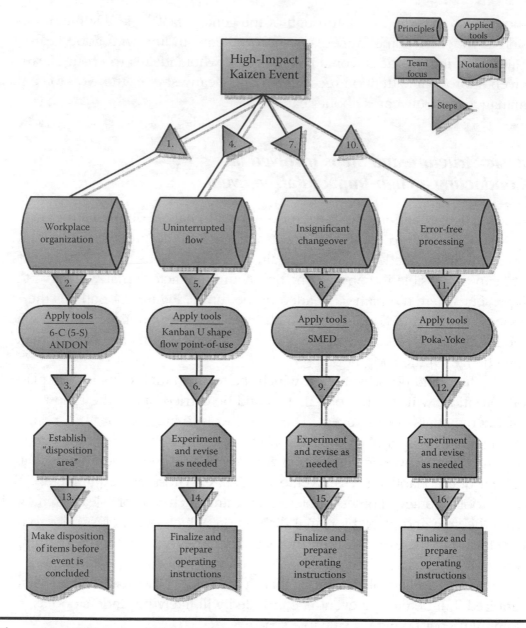

Figure 4.4 High-Impact Kaizen event.

applied, in order to get the best out of the event, the sequential steps outlined should be followed.

It should be noted that under uninterrupted flow, insignificant changeover, and error-free processing, the common notation spelled out is, "Experiment and revise as needed." This is meant to point out that participants assigned to these particular endeavors should not feel bound to getting it absolutely right the first time. In fact, they seldom will. A reasonable amount of freedom has to be given to doing things over during the course of the event, if there's a means to substantially improve the changes conducted. This, as mentioned, calls for maintenance to be fully responsive to doing the same basic kind of work over again, if called upon. Thus, maintenance personnel have to be some of the best-trained employees in Lean and appreciative of the fact that any work they are asked to do in support is far from being wasted effort.

Getting the Most Out of Training and Implementation Kaizen

A good deal of what was described for a high-impact Kaizen event applies to Training and Implementation (TI) Kaizen, but on a smaller scale. The purpose of a TI event is to train a relatively small number of employees (normally 10 to 15) in the basics of Lean Manufacturing and then to allow them to apply a portion of what they'd learned on the shop floor. The duration of the event is typically three days, a third of which is spent on classroom training and the rest on making change to a select process.

To get the most out of a TI event the selected production process should involve a line of equipment, a subassembly process, or the like. The application of Kaizen would typically include only one or two of the major tools involved, such as placing a focus on SMED (Single Minute Exchange of Dies) or Poka-Yoke (mistake proofing). But in every case, workplace organization and the 6-Cs should be a substantial part of the change conducted. This is because workplace organization is the foundation for continuous improvement. The idea isn't to make a full and complete change to the equipment and processing involved, although this sometimes happens depending on the participants' understanding and ability. The chief purpose is to use the exercise to allow the participants to see firsthand precisely how change should be made and the subsequent benefits involved.

Figure 4.5 is a window diagram of a typical TI event. The classroom training starts by providing a basic overview of Just-in-Time (JIT) manufacturing and exposure to the "seven deadly wastes" of conventional manufacturing. This is followed by a one-hour floor exercise that allows participants to explore and identify such wastes in the area in which they will be making change. Training is then provided on workplace organization and the principal tools they will be using (typically SMED, Poka-Yoke, or TPM). In the wrap-up session, the group, with the help of the Lean coordinator, collectively express to the plant manager and staff what they have learned and what their plan of action will be.

DAY ONE	DAY TWO	DAY THREE
INTRODUCTION	FLOOR WORK MAKING CHANGE TO THE SELECTED PRODUCTION PROCESS	FLOOR WORK CONTINUES
OVERVIEW OF JIT MANUFACTURING THE SEVEN DEADLY WASTES (HIDDEN WASTES)		
FLOOR EXERCISE: IDENTIFYING HIDDEN WASTES		
LUNCH	LUNCH	LUNCH
WORKPLACE ORGANIZATION	FLOOR WORK CONTINUES	FLOOR WORK CONTINUES
TRAINING IN SPECIFIC TOOL TO BE UTILIZED DURING THE EVENT: SMED, POKA-YOKE, OTHER		
WRAP UP: GROUP SHARES ITS LEARNING WITH MANAGEMENT AND SPEAKS TO PLANS FOR CHANGE	WRAP UP: GROUP REPORTS TO MANAGEMENT THE CHANGES MADE & LESSONS LEARNED	EVENT WRAP UP: GROUP REPORTS RESULTS & ANY FOLLOW-UP ACTION REQ'D—FOLLOWED BY PLANT TOUR

Figure 4.5 TI event window diagram.

The following two days are spent making change on the shop floor, which may indeed require some overtime by the participants involved. After lunch on the third day, one or two participants are chosen to work on a presentation of the results achieved, while the remaining participants continue to work on the shop floor as needed. The last hour of the day is spent on presenting the accomplishments made, followed by a plant tour to allow a firsthand look at results.

Although it is not absolutely essential to reassemble the participants of a TI event at some point in the future in order to return to the area and review what further progress has taken place, this has been proven to be a beneficial step. Doing so serves three purposes:

1. It gives the participants the feeling their work was important enough for management to allow them the time to regroup, review, and discuss the changes made, along with where other opportunities exist.
2. The production supervisor over the area should conduct the review, which leads to the second purpose. That purpose is for the leadership of the area to remain cognizant that if any slippage occurs they hold the responsibility of explaining why to the group that so diligently worked in making the change.
3. The final purpose is to raise the attention level of the general work-force to the fact that not only making change is extremely important, but keeping it intact and building on it is vital to the long-term success of the company.

Although some of the event and follow-up activity mentioned may sound a little unnecessary, companies cannot forget that the type of massive change required simply cannot be carried out unless actions are accompanied with a reasonable level of individual recognition and accountability. Here are some things to consider in accordance with that objective:

- Show strong support and appreciation for those who assume the responsibility to learn, apply, and make appropriate change on the shop floor and elsewhere, in keeping with good Lean Manufacturing practices.
- Consistently highlight such change in the company's newsletter and through other forms of posted and written communications.
- Provide small rewards for those who participate in making appropriate change. Nothing big. Something as simple as a hot dog and a Coke will usually suffice.
- Consider holding an annual luncheon for employees carrying a Lean Manufacturing theme. Make it an opportunity for the plant manager and others to express the importance of the process and provide the time for an employee or a number of employees who are enthused about the process to say a few words in support. One could be surprised at those willing to do so.

■ Consider highlighting and referring to the Lean Manufacturing effort being conducted as the company's new production system. *The Franklin Production System,* as an example. This personalizes the process and will typically elevate support and enthusiasm.

What may be coming to light is the fact that successfully implementing Lean Manufacturing goes much further than making change to production processes. It also requires changing a general mind-set that has been in place for years on end. This means correcting habits, changing perceptions and paradigms, and coming to see the business of manufacturing from a totally different perspective. Therefore, business as usual simply won't get the job done.

Training and implementation Kaizen is truly the backbone of a Lean Manufacturing initiative. It is the primary means of providing the needed knowledge to employees and carrying out ongoing change throughout the factory. It therefore deserves proper attention and a game plan that serves to utilize it to its fullest.

Modifying the Rules for the Purchase of New Equipment

Although a training and implementation Kaizen event is principally aimed at making the best use of existing equipment and facilities, there are occasions where the need for a new piece of production equipment becomes clearly obvious. Procurement of new equipment usually can't be accomplished over the normal course of a TI event, however, the event itself may be responsible for bringing new equipment aboard at some point. This in turn calls for some needed rule changes in both equipment design and procurement.

Production equipment comes in two basic forms: the first is equipment purchased as a stock item ready for use (a standard upright spot welder, for example), and the other is specially designed equipment to meet a specific production application. In both cases preprocurement procedures should include a "Lean Manufacturing equipment checklist" (see Exhibit 4.1). Every piece of stock item equipment, of course, will not meet all of the specifications noted on the form. However, the exercise of going over the checklist with the supplier should still be conducted. In some cases the supplier will make an effort to modify stock equipment to satisfy the needs spelled out on the form. On specially designed equipment, however, the supplier should be fully expected to produce the equipment with the specifications outlined in mind.

	Yes	No
1. Fast/easy maintenance built in	____	____
2. Easy to move/relocate	____	____
3. Excellent operating safety features	____	____
4. Requires minimum setup/changeover	____	____
5. Quick hook-up features built in	____	____
6. Low risk of air/hydraulic leaks	____	____

Note: The specific design of this document would include these items, along with other factors that are in keeping with good Lean Manufacturing principles and guidelines.

Exhibit 4.1   Example: Lean Manufacturing equipment checklist.

Planned Frequency of Training and Implementation Kaizen Events

The appropriate number of TI events will vary from company to company; but in larger operations the goal should be to have a minimum of 50% of the workforce trained the first year. For an operation with 400 employees (considering the size of the event will typically average 10 to 15 participants) this would mean around 15 events would have to be conducted the first year, averaging slightly more than one event each month. For smaller operations (100 employees or less) the goal should be to have all employees trained the first year, which would mean roughly one event a month until all employees are covered.

For the manufacturing operation wishing to assume a very aggressive training schedule, two events per month are typically all that can be scheduled and performed in an adequate manner. That would mean the maximum number of employees trained the first year would be roughly 200. Thus, for a firm with 400 employees the time span required to train everyone would approach two years in duration. However, there is no need for dismay. Making a full Lean transition in a large factory that's been driven for years with a batch mentality will take one to three years, depending on numerous factors. The important thing is to get started and press forward as aggressively as possible. Rest assured, however, it will not be two to three years before a company begins to see the benefits in terms of improved product quality, reduced operating expenses, greater overall flexibility, and steadily increasing profit margins.

Driving the Use of Problem Resolution Kaizen

Problem resolution Kaizen is unquestionably the least effectively used process in the Kaizen tool box. Most often addressing serious production issues that have an impact on throughput, operating costs, and the like are handled the old way. An example follows.

> Joe, the production supervisor, has a problem with a key piece of equipment. He calls Fred, the production engineering manager, for assistance. Joe and Fred end up going to the equipment and looking the situation over. Fred believes the problem boils down to the need for a more frequent tool change. Joe in turn contacts maintenance and the tool change is made. After the tool change everything appears to be fine so the issue is closed and production continues. When Joe arrives at work the following morning he finds a huge amount of inventory produced on the equipment during the second shift, which has to be scrapped or reworked. He finds himself immediately behind schedule and it becomes clear the fix didn't fully resolve the problem. In fact, if anything, it appeared to make it worse. Sound familiar?

The principal reason the problem went unresolved was because no one went to the trouble of clearly establishing the root cause. As a result this not only had a negative impact on meeting the schedule, but added waste in the form of scrap or rework. Getting down to the root cause and deciding on a permanent fix takes a conscious effort to pause long enough to see that it occurs.

Quick fixes commonly end up costing more time, energy, and effort than forcing one's self to indulge in a process aimed at correcting the problem permanently. That is the purpose of problem resolution Kaizen. Given this was used, Joe and Fred would have discovered that although an unscheduled tool change was indeed warranted, the actual root cause was machine wear on the drive shaft, creating a vibration that served to establish unusual and excessive tool wear. This in turn would mean that until the machine could be taken out of production and the drive shaft fully repaired, a revised schedule involving a more frequent tool change would need to be arranged.

A simple SPC (Statistical Process Control) chart on the equipment would have alerted the operator that the process was trending out of control, and would have allowed the operator to shut the equipment down before scrap and rework was produced, however, an SPC chart does not have the ability to say what was indeed creating the problem. Establishing and correcting the actual root cause requires an entirely different process.

Applying the Science of 5-W

Getting down to the root cause usually requires some persistent delving into the issue. A simple straightforward method commonly referred to as the "5-Whys" (or 5-W) involves asking *why* up to five (5) times. The following example is one to which almost anyone can relate.

Little Johnny comes home with a poor report card in English. His mother proceeds to sit him down and investigate. The conversation goes something like this.

Mother: "Johnny, why are your grades so poor in English?"
Little Johnny: "My teacher doesn't like me."
Mother: "Why doesn't your teacher like you?"
Little Johnny: "'Cause she likes the kids up front and Karen."
Mother: "Who's Karen?"
Little Johnny: "The girl that sits next to me, in the back."
Mother: "But why would she like Karen better?"
Little Johnny: "'Cause Karen can read the blackboard and I can't."

After the third *Why* it starts to become clear that the problem isn't that the teacher doesn't like little Johnny. The problem is that Johnny is having difficulty reading what the teacher writes on the blackboard. After a trip to the eye doctor, the lad gets a pair of glasses and his grades immediately begin to improve. Had the mother taken Johnny's first response without delving deeper she could have ended up making a trip to confront the teacher on Johnny's premise that she simply didn't like him, embarrassing herself and delaying a correction to the real problem.

Exhibit 4.2 outlines the benefits that a simple "why" can have. If it's used in the proper manner wisdom is gained, help is extended in establishing the root cause, and the process commonly yields thoughts and ideas that can often resolve the issue without going further. The proper manner of undertaking this approach, of course, is a polite respectful probing until some meaningful light is shed on the issue. The best way of approaching someone with the intention of using this technique is to tell them how the exercise works and what is hoped to be gained. In

W = Wisdom is gained.

H = Help is given in establishing root cause.

Y = Yield potentially provides a firm solution.

Exhibit 4.2   The inherent benefits of "Why?"

most cases they will willingly participate. However, walking up to someone and starting to pummel them repeatedly with "Why?" will only create unneeded frustration and can actually end up offending the very person who potentially has the most to offer.

Most people would wisely do what Johnny's mother did. However, very smart people often forget to apply the same logic they use in their personal lives after they walk through the doors of the place they're employed. Some of this is due to the perception they should check their brains at the door and do what they're told. This is especially true of the average production workers, who frequently gain the feeling that all the company is truly interested in is their staying at their machines, keeping their mouths shut, and running production. Although the advent of Lean Manufacturing has changed that perception somewhat, on average there's still a great deal of room for improvement.

Driving the use of problem resolution Kaizen typically boils down to management insisting the root cause must be established and reported on for any problem elevated to the top. This means the first thing out of a production or plant manager's mouth in response to a noted problem should be, "What's the root cause?" This says to subordinates they are free to come with problems but they hold an obligation to strive to clearly understand what caused that problem and ideally what can be done to correct it. When a production problem occurs on a repetitive basis, a truthful answer to the question may indeed be that no one knows for certain. When this occurs, a problem resolution Kaizen event is the best way of addressing and permanently resolving the issue.

How to Conduct a Problem Resolution Event

There's an old saying that two heads are better than one. Leaving the fix to a repetitive production problem up to a single individual, regardless of how talented, doesn't negate the fact that if the right people are involved, two minds or more are better than one and will almost always arrive at the better solution. Although this isn't news to most managers, the sad fact is this typically doesn't happen. The common approach is to make certain the problem and others are covered with enough inventory—or enough leeway in the established manufacturing lead-time—to overcome (essentially hide) the real source of the problem. This doesn't happen because anyone is averse to fully resolving the problem, but because people are reluctant to slow the process of business as usual in order to do so. The old paradigm, "It isn't my job," needs to be thoroughly erased in manufacturing firms across America. Addressing serious production problems is everyone's job, especially if they have anything of importance to add and are called upon to participate.

The best manner of going about conducting a PRK event is to involve the following personnel as direct participants:

- The production engineer responsible for performing sustaining work on the equipment or production processing involved
- One or two of the more knowledgeable operators involved
- Someone from the quality assurance function
- The supervisor over the area involved
- A representative from the maintenance function
- If possible, the original equipment manufacturer or equipment supplier

The Lean Manufacturing coordinator should be called on to facilitate the effort, because he should be the most knowledgeable in the application of the appropriate tools the group will use in getting down to the root cause and in assuring that the agreed-upon fix meets good Lean Manufacturing practices. Figure 4.6 is a window diagram of a typical PRK event.

The event starts with an overview by the area supervisor regarding the problem and the impact it has had on achieving production, along with any history associated with trying to resolve the problem and how the matter has generally been handled in the past. The next thing is for the production engineer to explain as much as possible about equipment and processing specifications, general maintenance procedures, the established operating methods, and any personal experiences in striving to address the problem. This is followed by the Lean coordinator conducting training in Poka-Yoke or, in the case of the group having had previous exposure to the training, a short refresher in Poka-Yoke. The group then collectively descends to the shop floor to observe and record what is actually occurring, noting:

DAY ONE	DAY TWO
INTRODUCTION PROBLEM I.D. AND DISCUSSION	FURTHER FLOOR WORK TO ESTABLISH ROOT CAUSE AND PUT A RELIABLE FIX IN PLACE
TRAINING / REFRESHER IN 5-W, POKA-YOKE, 4 PRINCIPLES OF WASTE-FREE MANUFACTURING	
FLOOR WORK EXAMINING PROCESS	
LUNCH	LUNCH
FLOOR WORK CONTINUES	FLOOR WORK CONTINUES
WRAP-UP SESSION: GROUP DISCUSSES FINDINGS AND PROSPECTS TO RESOLVING ISSUE	WRAP-UP SESSION: GROUP REPORTS TO MANAGEMENT THE CHANGES MADE AND LESSONS LEARNED

Figure 4.6 PRK event window diagram.

- The methods being used.
- Any downtime that might occur.
- The general production flow.
- The quality control procedures utilized.
- How the parts are handled, transferred, and stored.
- Any other observations noteworthy in getting down to the root cause.
- Note: Out of this will usually come a number of thoughts and ideas as to how to resolve the problem, but nothing should be taken for granted at this point.

In conducting the floor exercise someone in the group should perform a 5-W with the operators involved. This often leads to the discovery of things that may not be apparent at the particular time the group is on the shop floor.

At the end of the day the group reassembles in the meeting room and shares its findings with general management. From there, a number of things can happen. It could be that a need arises to bring the original equipment manufacturer or supplier in to work with the production engineer and maintenance representative, in seeing that the equipment is properly meeting established specifications and otherwise obtain the benefit of their input.

Typically, when a serious production problem occurs on a repeated basis, the odds are high that bringing in the equipment manufacturer or supplier has already happened. But asking them to join the event on the front end is the ideal thing to do, if a commitment can be achieved.

The first half of the second day is spent on the shop floor repeating the exercise performed the day before. Generally, additional things will be learned and the fix will become apparent. The group reassembles after lunch and comes to an agreement on a permanent solution to the problem. In some cases the fix will be simple enough to incorporate before the first day of the event is concluded. In other situations the fix may be clear, but it could take time to fully implement. Either way, the event is concluded with a short presentation to management regarding findings and the fix involved. In the case of requiring additional time to implement, after the solution is finally implemented the group reassembles for a short tour, in order to review the fix and to bring the matter to a conclusion.

With reference to the pie chart in Chapter 2 (Figure 2.1), problem resolution Kaizen does not normally equate to a large percentage of the overall accomplishments made in advancing Lean (roughly 5% of the total). But it is important to note what will generally come out of the PRK events are some obvious opportunities for further improvements that are in line with good Lean Manufacturing practices. Thus, a PRK event can serve more than a single purpose. It can heighten awareness and become just another vital step in the right direction.

Essential Tools Utilized in a Problem Resolution Event

Outside of the use of the 5-W, the application of Poka-Yoke is typically called for. This is especially true if the problem is equipment related. In situations where the equipment doesn't end up being the culprit, there should still be an effort made to take advantage of the opportunity and extend a healthy dose of mistake proofing. In addition, TPM should be evaluated and good Workplace Organization should be established. The principal task is to resolve the problem the group assembled to address. But inasmuch as they have been assembled and are using Kaizen to resolve the matter, as much work as possible should be extended in advancing good Lean Manufacturing practices.

As mentioned in the Introduction, the how-to involved in using and applying the prescribed tools noted, most of which were derived from the Toyota Production System, isn't specifically covered herein. However, there are literally hundreds of books and other published works related to the use of the tools, the techniques involved, and how to go about conducting training in their use. A list considered to be some of the better material available can be found in Appendix A.

Keeping the Principles of Uninterrupted Flow and Workplace Organization in Mind

Something that should always be kept in mind when striving to resolve production problems and issues, whether they are quality or throughput related, is where the application of continuous flow techniques fit. The fewer steps involved and the least amount of disruption in flow due to storage, inspection, and other factors, the greater chance errors can be avoided. Figure 4.7 points out the importance of this consideration, using a machine shop example. When material and components can bypass common receiving and inspection steps in the process—through establishing confidence in the supplier with vendor certification, along with laying out the production processing involved to avoid common stocking arrangements before and after production and by making operators responsible for inspection—the easier it becomes to address and fully resolve problems and issues. In addition, some superb steps are

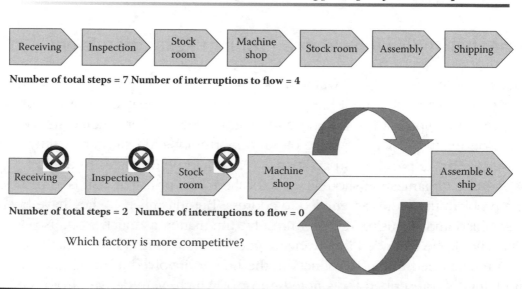

Example of time compression using principle of Uninterrupted Flow

Number of total steps = 7 Number of interruptions to flow = 4

Number of total steps = 2   Number of interruptions to flow = 0

Which factory is more competitive?

Figure 4.7   Example of time compression (Uninterrupted Flow).

made in time compression and improving operational flexibility. The question at the bottom of Figure 4.7 is: "Which factory is more competitive?" The obvious answer, of course, is the one with the least amount of disruptions in flow.

Along with a focus on uninterrupted flow, the principle of workplace organization should also be taken into consideration and applied by the group as time allows. Very often, repetitive production problems boil down to a simple lack of operator instructions (visual controls), which good workplace organization can most effectively serve to address.

Understanding the Role and Scope of Sustaining Kaizen

Sustaining Kaizen consists of two key components. The first is Kaizen activity carried out at an individual job level. This usually requires the implementation of an incentive, such as that outlined for a WRAP initiative, in order to strongly encourage employee participation, both on the shop floor and in the various office functions involved. The second component of Sustaining Kaizen (SK) is a formal Kaizen event aimed at further enhancing a change that was made or expanding the scope of an established change to other areas of the office or factory.

In most cases the participants of an SK event would consist of some employees involved in making the original change, along with representatives from the area where the change is due to be imposed. The latter participants may or may not have had specific Lean and Kaizen training, which this particular event is not designed to achieve, outside of a refresher on the basics for everyone. Therefore, there could be some involved who are asked to participate on the basis of trust in the process and the merits of the initial change made.

Figure 4.8 provides a visual overview of Sustaining Kaizen. The combination of individual job improvements and formal SK events combine to make sustained advancements in Lean Manufacturing, which to a large degree is a never-ending process for a factory.

The fundamental principles of Waste-Free Manufacturing become the guiding values for every action taken. Again, those principles are workplace organization, uninterrupted flow, insignificant changeover, and error-free processing. Participants are asked throughout an SK event where the principles apply to any change being considered. If a suggested change doesn't clearly fit within the scope of one or more of the four guiding principles, it is considered suspect and care is taken before it is implemented. In the

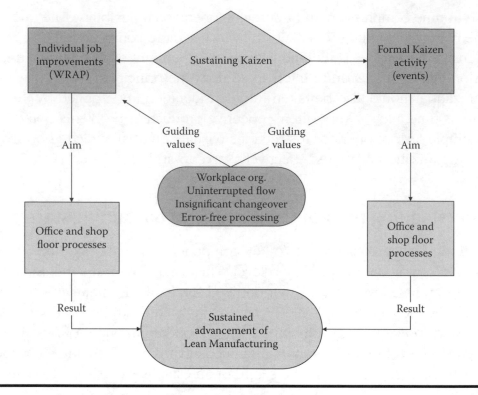

Figure 4.8 The components of Sustaining Kaizen.

individual job improvement arena, a WRAP initiative is spelled out. Although this isn't absolutely required, it serves to greatly enhance the process. In fact, without something similar, the chance of getting this important segment of sustaining Kaizen off the ground is limited, at best.

The basic role of sustaining Kaizen is to maintain a focus on enhancing improvements to previous changes made and, to every extent possible, to make Kaizen a daily activity. The intended scope covers both shop floor and office processes and the use of sustaining Kaizen involves both formal and informal Kaizen activity. One of the best ways to define the critical importance of sustaining Kaizen is to view it as the glue that holds the entire process of Kaizen together. Without it, the very best work with other forms of Kaizen will tend to erode over time.

Something to keep in mind is the fact that "niceties" often arise in Kaizen-related activities. This is especially true in the area of individual job improvements. For example, someone thinks a new workstation PC with more features would be nice. However, if it does nothing in advancing one the four principles the thought should be politely dismissed. Keeping a proper focus on the mission at hand, which is to improve operational efficiency, has to remain at the forefront of all Kaizen-related activity.

Implementing a WRAP Initiative

A thoughtfully established and well-run WRAP initiative can be an extremely positive influence on getting the absolute most out of Kaizen. In fact, it is essential in making Kaizen a formidable competitive weapon. The following address some important things to keep in mind in approaching the task.

When to Start a WRAP Initiative

The general workforce should have exposure to Lean before a WRAP initiative is undertaken. For companies just starting a Lean effort, WRAP should not be considered until employee training has been conducted in the fundamentals of Lean Manufacturing, and some level of Lean-oriented change has been made on the shop floor. A good guideline for when to consider a WRAP program would be 8 to 12 months after a Lean initiative was underway, of course depending on the aggressiveness of workforce training. Providing the entire workforce with essential training is not only important to the overall success of Lean Manufacturing, but critical to a WRAP initiative. However, the planning phase should begin at the earliest possible date.

Planning Phase Considerations

There are three important considerations to keep in mind. The first has to do with how a WRAP program will be measured for remuneration. As previously noted, complicated formulas should be avoided. When establishing incentives of any kind almost every business function wants a piece of the action, so to speak. Quality assurance may vie for a number of standard quality measurements to be included in the calculation. Accounting may want the actual cost of scrap and rework measured against targeted budgets or forecasts, and the list could go on. These should be avoided, because it's important to keep things simple and straightforward when it comes to an incentive for making on-the-job improvements through the use of Kaizen.

One of the best ways to decide if a change is qualified for a bonus is to evaluate it on the basis of the four guiding principles of waste-free manufacturing. If it meets those guidelines it is an acceptable improvement and no further justification is warranted. There is, however, one exception. If the change requires the purchase of new equipment, tools, and so on (something that can't be built with common material the maintenance function carries as stock), the idea is recorded and tabled for further consideration. Once a year, a WRAP board, appointed and headed by the plant manager, assembles to review any tabled ideas and to make a firm decision if the expenditure can be justified. If it can, the employee is allowed to take the idea to a conclusion and once fully accomplished, the bonus is paid.

This leads to the point that WRAP is not a "pay-for-an-idea" incentive. An idea is only as good as the employee's ability, coupled with the assistance of the area supervisor, to fully implement. Employees must come to understand that the chief purpose of Kaizen is to make improvements to the equipment and facilities on hand. However, anything that can be readily built by maintenance to assist in making an improvement, such as special fixtures and the like, should be deemed acceptable. There is a bit of a fine line in judgment required when it comes to special raw materials to accommodate a change. Good common sense should prevail, thus all the more reason for establishing a well-thought out and management-approved budget for Kaizen on the front end of a Lean initiative.

The second important planning consideration is what level of compensation will be paid for a successfully applied change. There are obviously a number of ways to approach this, but a simple option is a flat bonus of $50 for accepted improvements. Regardless of the decision made on the amount of compensation, a fund should be established for WRAP and included in the Lean coordinator's operating budget. The total annual sum will vary from factory to factory, depending on numerous factors such as the amount of bonus paid per improvement, the number of employees (both hourly and salaried), the level of participation, and so on.

In estimating the initial budget for WRAP, 25 to 30% workforce participation is a reasonable number to use, especially at the offset. I was a party to putting such a program in place in a factory that had over 800 employees. During the first full year 27% of the workforce participated with accepted improvements. Inasmuch as that 27% participation occurred under a rather aggressive application of Lean, it should be safe to use a factor of 25% in estimating a factory's initial budget for WRAP. Using this, the formula would be: Total # of employees [hourly and salaried] × .25 × bonus paid per accepted improvement = estimated annual budgeted amount. For a factory with 600 employees, applying a flat $50 bonus per accepted improvement, the estimated annual budget for WRAP would be: $600 \times .25 \times \$50 = \$7500$. Again, the basic guideline for acceptance would be: Does the change do anything to enhance workplace organization, uninterrupted flow, insignificant changeover, or error-free processing? If it does, it should be viewed as an accepted improvement. If it doesn't, it simply doesn't apply.

One last thing to address in the planning phase is the depth and repetitiveness of an accepted change. An issue that will most likely arise is how great a change warrants a paid bonus. For example, an operator might suggest to his supervisor that a simple hanging device for an air driver would enhance workplace organization. The fix would be quick and easy and the question in someone's mind could be, does this warrant the same level of compensation (applied bonus) as something that has a great deal more impact on the overall operation. The answer is yes. The value of an improvement can't always be measured or viewed in terms of strict payback. It has to be seen as an investment in training employees and changing the mind-set of the workforce. In most cases, however, improvements will not fall into this category.

Another question that can arise involves a repetitive improvement. For example, someone in one area of the factory sees a change made that earns a bonus and decides to use that as an accepted improvement in his area. Should a bonus be paid for essentially the same idea generated elsewhere in the factory? The answer is no. The first employee coming up with the idea is the only one rewarded, regardless of when, where, and to what extent that improvement is later used somewhere else in the factory.

A WRAP initiative is aimed at not only generating ideas for improvements at an individual job level, but recognizing and rewarding those who are creative and take it upon themselves to step forward in making change for the better. Unfortunately, there will be some in the workforce who never meet the established standard for an accepted improvement. This doesn't necessarily mean they are inept or disinterested. What it does mean, is that the supervisor has to accept the challenge of providing them the personal time and attention they may need to become proficient at making change for the better in their jobs. As expressed earlier, the role of the supervisor has to change from being purely directional (demanding) in nature, to being more inspirational (motivational) in practice. And it is additionally why shop floor supervisors should carry some reasonably heavily weighted objectives aimed at bringing the workforce along.

A man I very highly admired once told me that any kind of special incentive aimed at urging employees to "do something they should be doing anyway" set the wrong precedence. I wholeheartedly agree, with one exception. The wastes we have in conventional manufacturing have come about because management decided to train employees in a specific way of doing business. With a Lean Manufacturing initiative we are essentially saying to them to forget what they were taught in the past and do their job in an entirely different manner. We further complicate the issue by asking them to work inside the parameters of the old system of production, while such an effort is underway; and last but not least, we're suddenly asking them for ideas after years of leaving the impression the only thing we were interested in was for them to stay busy. Rewarding employees with a reasonably decent incentive for good ideas that serves to advance the full implementation of Lean is the least we can do in working ourselves out of the muck and mire of conventional manufacturing practices, which are there because industry chose a wasteful system of production.

Typical Hurdles to Clear

The single biggest hurdle to clear in getting a WRAP initiative underway has to do with proven payback on investment. Typical cost justification measurements will simply come up lacking. There has to be a serious conviction for the type of change needed and many ideas for on-the-job improvements will not immediately show a return based on common justification standards.

Addressing this hurdle is where the F alliance noted in Chapter 2 (Figure 2.2) comes into play. The plant manager has to carry a reasonable amount of faith in the process and must help to clear this hurdle by insisting, if for no other reason, the budget and expenditures for a WRAP initiative should be a discretionary fund that can be used to enhance overall plant operations. The financial obligation would be to strive to live within the assigned discretionary budget. Where the need arises for added funding, as a result of increased participation by the workforce, justification should be based on an overview of the changes being made and the overall impact they have had on such things as inventory, space reductions, manufacturing lead-time improvements and such, which aren't always easy to measure under typical cost justification standards, but that are certainly the proper thing to do.

Another potential hurdle is ensuring that each and every employee has a level starting field before a WRAP initiative is put into effect. This means the company has to take on both the obligation and expense of providing adequate training in the basics of Kaizen and Lean Manufacturing for each and every employee. This can be one of the more sizable costs associated with implementing Lean, but as noted the payback in terms of improvements made while the training is being conducted can be substantial. Depending on the size of the workforce, this training obligation could take a year (or slightly more) in the making. A word of caution, however, is not to shortcut the training with a four-hour overview of Lean Manufacturing or something similar. Employees not only need classroom training, but the chance to apply that training on the shop floor, which is precisely what a Training and Implementation Kaizen event is structured to do.

Summary Overview of the Process

Figure 4.9 provides a visual of the rollout of a WRAP initiative. It starts with phase #1 planning and is followed with phase #2 employee training (which is the longest task in duration) and finally a phase #3 communication and rollout phase, which are separate tasks but overlap to some degree. Figure 4.10 indicates the twofold objective of WRAP (coupling individual job improvement and the full insertion of Lean Manufacturing) and the keys to each individual objective outlined.

Communicating and Tracking Results

The matter of communicating and tracking results cannot be overemphasized. Keeping the attention at a maximum level is essential to success. The standard obligations of meeting production schedules, addressing problems that arise, and achieving established budgets and forecasts can seriously distract from a Lean/Kaizen initiative. Keeping a watchful eye on communications and overall results not only keeps the attention level of the workforce where it needs to be, but by nature of the activity itself helps to keep management focused on the mission.

A standard monthly company newsletter is certainly a good means of communicating and tracking the results of a Lean initiative, along with encouraging participation. But this should be reinforced with at least two biannual communication sessions by the plant manager, aimed exclusively at discussing the value of Lean, the progress that has been made, and the job yet to be done. Nothing carries more weight and influence. But some other communication and tracking opportunities are:

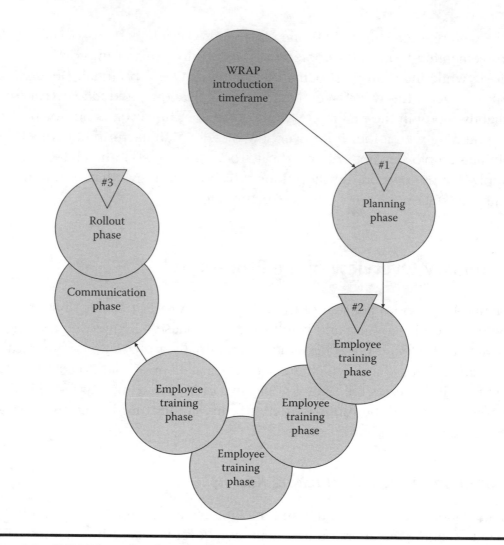

Figure 4.9 WRAP introduction timeframe.

■ Hold an annual Lean Manufacturing banquet aimed at celebrating overall accomplishments and recognizing those who have made individual contributions.

■ Use the local news media (newspaper, radio, or television) to communicate the effort being conducted and the kind of results being achieved. Most local news media take an interest in featuring an article or an interview on how a company is striving to remain competitive in today's world, if they are asked.

■ Keeping a "Lean Board" in an area of the factory that most employees visit, such as in or just outside the company cafeteria or break area. The purpose of the board is to indicate established goals and objectives,

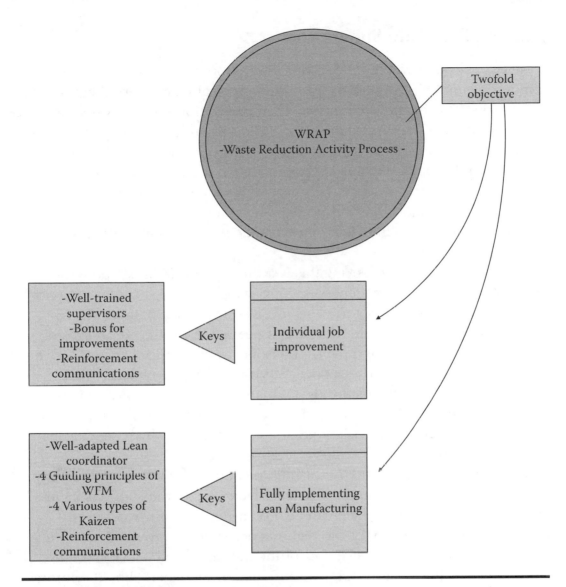

Figure 4.10 Twofold objectives of WRAP.

along with results and accomplishments, and various "words of wisdom" to encourage and motivate participation.

These are only a few of the numerous ways to work at communicating and tracking the progress of a Lean initiative. The point is to strive to keep the attention level as high as possible at all times. It's important to remember the task of Lean implementation is massive in nature and will not be accomplished if it is in any way felt by the workforce that it plays second fiddle to something else the plant is doing.

Training First-Line Supervisors

If there is one area where the least effective effort has been made when it comes to Lean, it boils down to the training typically given first-line floor supervisors. This happens because normal production schedules have to be achieved while a Lean Manufacturing effort is underway. Key hourly employees are regularly excused to attend Kaizen events, however, the supervisor most often remains in the production area to ensure things are accomplished in their absence. This is understandable, but points to the fact that a special training effort needs to be extended exclusively to shop floor supervisors, who in essence are first-line management in the eyes of employees.

The best method of achieving this is a three-phase training approach. The first phase is Lean awareness training, which ideally is conducted prior to introducing Lean to a factory. This involves a one-day training session that can be conducted on the weekend if needed. Here the advantages of Lean are fully explained and the supervisors are given the chance to work as a team in identifying "The Hidden Wastes" (see my book, *Fast Track to Waste-Free Manufacturing* for more detail on the hidden wastes) in a selected area of the factory, which doesn't have to occur while production is underway, but serves to clearly point out inefficiencies with the existing system of production.

The second phase of supervisor training should be conducted directly after the plant's first high-impact Kaizen event. Here, the supervisors are brought together again (which can also be arranged for a weekend if needed) and an overview of the high-impact event is provided for their review, including an extensive tour of the area involved. In closing, the plant manager (or at least, the plant production manager) should reinforce the review by informing the supervisors that it is the intention of management to spread the same kind of change through the entire factory and that their support in the effort is not only expected but anticipated.

The third phase of the training is to strive to have supervisors cover for each other so over a period of time each and every shop floor supervisor can attend and participate in at least one formal Kaizen event.

Key Summary Points

The overall scope of Kaizen activity:

■ Utilizing Kaizen to its fullest encompasses more than a single-minded process. It is instead a series of established activities that have different purposes and lead to different results, all of which are aimed at fully and effectively inserting the principles of Lean throughout an entire business enterprise (see "Progressive Kaizen Tool Box").

How "WRAP" fits the picture:

■ In most Lean initiatives an effort is made to create the understanding that Kaizen means continuous improvement. Although that is true in the sense of a broad translation, it simply doesn't carry the inspirational impact of "waste reduction." Couple this with "activity" and "process" and you have "Waste Reduction Activity Process" (WRAP) which clearly serves to address what the effort is aimed at accomplishing, "wrapping" the implementation of Lean in an active and ongoing waste reduction effort, aimed at making sound continuous improvement (see "Advantages of Labeling Kaizen Activity 'Waste Reduction'").

A factory's first "pull" zone:

■ An important factor that should be considered above anything else is where a viable "pull-production" process cannot only be implemented, but will stand as the leading example of where the factory is headed in the future. With this in mind, one or more lines in a plant's final assembly area become the most ideal spot to start (see "The Value of Putting the First 'Pull Zone' in Final Assembly").

A factory's first high-impact event:

■ A factory's first high-impact Kaizen event will in most cases set the stage for precisely where an operation is headed in making change for the better. It therefore requires a good deal of forethought as to precisely what area of the factory will be the recipient of such change and who the participants for such an event will be. The duration of the event varies from one to two weeks, depending on the area involved and the depth of penetration set forth for the event (see "Conducting the Factory's First High-Impact Kaizen Event").

The issue of problem resolution:

■ Quick fixes commonly end up costing more time, energy, and effort than forcing one's self to indulge in a process aimed at correcting the problem once and for all. That is the expressed purpose of problem resolution Kaizen (see "Driving the Use of Problem Resolution Kaizen").

What a WRAP initiative is and isn't:

■ A WRAP initiative is not a "pay-for-an-idea" incentive. An idea is only as good as the employee's ability (coupled with the assistance of the area supervisor) to fully implement it. There will additionally be occasions where an idea is good but requires the expense of new equipment. Employees must come to understand one of the chief purposes of Kaizen is to make improvements to the equipment and facilities on hand (see "Implementing a WRAP Initiative").

Chapter 5

Other Key Facets of Getting the Most Out of Kaizen

The purpose of this chapter is to address a number of items that haven't been fully covered, but which can aid tremendously in enhancing a Progressive Kaizen effort and the full implementation of Lean Manufacturing.

Advancing the Role of Owner-Operator to "Lean Equipment Specialist"

In Chapter 2 the need for the consideration of owner-operators was addressed. The topic is something I've never failed to cover in any of my previous writings because it can be one of the most helpful steps an operation can take in keeping its equipment and production processing running effectively, especially key production equipment. However, experience has taught if the role of the owner-operator is expanded to include an ongoing focus on good Lean Manufacturing principles as a daily function of the job, even further progress can be made. I've come to refer to this elevation of duties as the "Lean Equipment (LE) specialist."

It is important to note there are basically three types of production equipment found in almost every manufacturing operation. Class I, which involves equipment and machines that are essentially "shelf items" which are purchased, installed, and utilized with no modification (a basic upright spot welder, a standard lathe, etc.). Class II equipment involves the same type outlined for Class I but consists of machines that have been modified somewhat to meet specific processing requirements. Class III equipment is uniquely designed machines, typically built from the ground up, and essentially one of a kind. Some Class II and all Class III equipment generically fit the category of "key equipment."

"Key" equipment is just that; equipment key to making and delivering the product a factory was designed to produce. Although it isn't something most businesses would desire to do, the work performed on standard shelf-item equipment could be outsourced if the need arose. That simply isn't the case with some Class II and most all Class III production equipment. These essentially become the lifeline of the business and if and when they fail to operate as intended—or fail to operate at all—a business is in serious jeopardy. That is why a special focus placed on engineering key production equipment to support a Lean effort has been outlined as being critical to implementation. But if done right, the effort fits a twofold purpose: to have equipment that supports the principles of Lean and to provide added assurance that equipment is always capable of running (producing), when and as production requirements call for them.

The role of the LE specialist is designed to fit both of these needs. Establishing the role begins with required training under the classification of an LE apprentice, and it is only after displaying the full ability to do the job that he or she is elevated to the position of a LE specialist. To expound on the classification, the following areas of address are covered:

1. Where LE specialists should be considered
2. The basic pay structure for such a classification
3. The percentage of the workforce that should hold the classification
4. The specific training involved for elevation to a LE specialist

Where LE Specialists Should Be Considered

Again, every manufacturing operation has key production equipment, whether it is recognized as such or not. In the form of definition, the term *key* applies to equipment that is highly specialized, very often complex in nature, and usually engineered to perform a unique production task. In most manufacturing operations typically about 25% or less of all factory production equipment would fit the category, but there are situations where a higher percentage applies. Outside of the type of things most companies do in assuring the uptime capability of their equipment, such as the application of TPM, the need exists to provide an added level of equipment dependability where possible. One manner of supporting this is through the use of LE specialists.

A Lean equipment specialist is an employee who has proven expertise in operating the equipment involved and has been given the added responsibility for assuring the quality of the parts produced, along with general maintenance and equipment upkeep. In one respect they can be viewed somewhat like a first sergeant in the Army. Although they are not the commander in charge of the unit, when it comes to action on the field they assume an unquestionable authority. The LE specialist effectively owns the performance of the equipment and holds an above-average responsibility to ensure what comes off that equipment fits the need of the next user in the process. Along with good component quality and machine upkeep, they understand and accept responsibility for:

1. Ensuring they are producing to the next user's requirements, in terms of both the quantity of parts produced and any special part orientation requirements, related to the stacking and transfer of parts before they are released for delivery. The LE specialist personally determines what such arrangements should be by speaking with the supervisor of the receiving department or production zone, as well as the operators involved. They then proceed to pass the information on to the sustaining engineer so it can be included in formal "routing sheets" and other processing documentation.
2. Making appropriate disposition of parts deemed unacceptable. Although this would be a rare to nonexistent occurrence under good operating procedures and especially where mistake proofing has been applied, Lean equipment specialists understand the job classification carries the obligation to ensure nothing but fully acceptable parts are delivered to the next user in the chain. If for some reason this isn't the case, they hold the responsibility of personally applying themselves to correcting the situation.
3. Understanding good Lean principles and applying those where opportunities exist, by passing on suggestions to the shop floor supervisor or the assigned production engineer for the area and in some cases simply taking it upon themselves to make the improvement without direct assistance.

Pay Structure for a Lean Equipment Specialist Classification

The pay grade of LE specialists will vary depending on numerous factors. However, it should be at least equal to the highest-paid direct labor position currently held in the factory. In many cases, the leading direct labor pay grade would be a "group leader." Group leaders typically possess the skill and experience to perform any of the operations in a specified area, along with holding the responsibility for training new operators and conducting fill-in duties in someone's absence. But regardless of which classification holds the honor of the highest pay grade, the LE specialist should be considered as having achieved the ultimate level of accomplishment on the factory floor. In accordance, the pay grade for the position should be reflective of that, because the LE specialist would be fully qualified to:

- Set up and change over the equipment as needed.
- Use any related quality assurance instruments and make certain they remain officially certified.
- Read and assess blueprints as needed.
- Perform on-the-job preventive maintenance.
- Understand the equipment specifications, limits, and so on.
- Recognize the advent of developing problems and take corrective action.
- Hold and utilize skills in SMED, Poka-Yoke, TPM, and workplace organization.

Percentage of Workforce Holding the Classification

The percentage of the workforce holding the classification of LE specialist should be reflective of the amount of equipment classified as "key." This requires a conscious effort by the company to identify clearly any equipment that meets the criteria for "key." To provide an example, assume a factory has 110 employees and 75 different pieces of production equipment. Out of that 75, twenty pieces of equipment have been identified as "key." The ratio of LE specialists in this particular case would be 18% of the workforce. (20/110 = 18%).

A word of caution applies, however. When a LE specialist ends up being equal to the highest-paid direct labor classification in the factory, experience has shown there will often be efforts extended to somehow make other operations "fit" that description. It's an easy trap to fall into and the entire pay structure for direct labor work can be negatively affected if appropriate care isn't taken.

A lengthy explanation about the difference between direct and indirect labor classifications probably isn't warranted, however, a brief description of the difference is that direct labor classifications involve work directly associated with building and assembling the products sold to the customer. Indirect labor classifications involve work for such things as material handling, inspection, and setup, which are "non-value-added" and should ideally be targeted for extinction under a good Lean Manufacturing effort.

Lean Equipment Apprentice Training

The training aspects for those selected for an apprentice role are covered in Figure 5.1. The training involved is normally a three-day exercise. To begin, potential LE apprentices are given basic training in Lean and the four guiding principles. This is followed by instructions in print reading and a more detailed level of training in TPM. The second day begins with an overview of the "Hidden Wastes" (built around Toyota's "Seven Deadly Wastes") followed by a factory floor exercise aimed at identifying such wastes in a select area of the factory. Due to the importance of the subject, TPM is "revisited" and special training is given in SMED. The last day starts with training in Poka-Yoke, followed by a classroom exercise exploring where opportunities for SMED and Poka-Yoke exist on the assigned equipment involved.

The afternoon session of the third day begins with a review of the entire training and the participants are given time to take notes, ask questions, and

DAY ONE	DAY TWO	DAY THREE
INTRODUCTION	EXPLORING THE HIDDEN WASTES	POKA-YOKE
LEAN MANUFACTURING AND THE FOUR GUIDING PRINCIPLES: • WORKPLACE ORG. • UNINTERRUPTED FLOW • INSIGNIFICANT CHANGEOVER • ERROR-FREE PROCESSING	FLOOR EXERCISE: IDENTIFYING THE HIDDEN WASTES	CLASSROOM EXERCISE: OPPORTUNITIES FOR POKA-YOKE AND SMED ON EQUIPMENT UTILIZED
LUNCH	LUNCH	LUNCH
PRINT READING	TPM REVISITED	STUDY PERIOD FOR FINAL EXAM
TPM	SMED	FINAL EXAM
WRAP-UP	WRAP-UP	WRAP-UP

Figure 5.1 Owner-operator training window diagram.

study for a final exam. The exam is given late in the afternoon and the training is wrapped up. The participants are informed of the results of the exam in a one-on-one meeting with the instructor the following day. Participants who pass the exam are given a certificate of completion and awarded the classification of LE apprentice. If a participant fails the exam (which is rare) but wishes to give it another try, he is rescheduled to take the training at the next available session. If the participant fails the exam a second time, however, he is disqualified from further pursuit of the position for a period of up to 12 months.

Certified LE apprentices are given a three-month period of observation by the Lean/Kaizen coordinator, along with an assigned task in keeping with the training received, which can vary greatly from one piece of equipment to another. But the end result should be an easily identified improvement involving SMED, Poka-Yoke, workplace organization, and the like. After completion of the trial period noted, along with the successful accomplishment of the assigned task, the operator is reclassified as an LE specialist and given an increase in base wage.

It's important to note that the operators involved are not rewarded with an increase in pay by simply applying and taking training to become LE specialists. They have to extend the personal effort required to show their ability before they are officially reclassified and awarded an increase. They additionally have to go through a refresher course once a year.

RELATED EXPERIENCE: The factory where I had my first opportunity at leading a Lean implementation effort became a huge learning experience for both me and those who so energetically worked at making it happen. We unquestionably made mistakes along the way, but in the end a remarkable shift was made in the manufacturing practices utilized.

Approximately six months into the effort, I was approached one morning by the production manager, Jim Wainwright, who told me he'd like to consider establishing a special labor classification for a select group of operators who had taken it upon themselves to do something special in support of the type of change we were trying to incorporate.

My first reaction wasn't totally in favor of the idea, feeling it gave the appearance of providing special rewards for the type of activity we were striving to drive through the entire rank and file as a standard part of doing the job. But Jim convinced me it was a case of more than that. Not willing to give up easily, he asked if I would accompany him to a key piece of equipment we had on one of our primary assembly lines. As we approached the machine, which had long been a source of nagging downtime problems, I recall remarking, "Going to see Old Faithful, huh?"

The machine had the function of producing a component subassembly, by automatically feeding parts into a large carousel, where they were positioned, coupled, and joined before being fed onto the main line. The machine had long been the source of persistent downtime problems due to lubricating oil and grime forming inside the carousel and affecting the correct orientation of parts, along with the joining process itself. Best described, it was a persistent nightmare, born out of an engineering effort to reduce direct labor requirements.

When we finally arrived Jim asked the operator to show me what he'd done. The operator proceeded to open the carousel and what I saw left me literally astonished. The inside was absolutely spotless! Jim went on to point out the operator had taken it upon himself to have maintenance install a simple air line that blew the overflow of lubricating oil away from the connecting mechanism into a special designed circular pan, that could be removed and emptied as needed. Jim said the operator had also taken it upon himself to personally clean the inside of the carousel twice daily, at the beginning and end of each shift, noting that since starting the practice the machine hadn't been down even once.

It became easy to see the point Jim was striving to make and the experience became the catalyst for our establishing an owner-operator classification. After a number of years in consulting, I saw an advantage to expanding the role to be further in keeping with advancing Lean, and went on to develop an outline for a LE specialist classification that applied to those who were not only proficient at running key factory equipment and who held the necessary skills required to perform preventive maintenance and other equipment sustaining duties, but also held the knowledge inherent to advancing good Lean Manufacturing practices.

Small things can sometimes make a big difference, things which production engineers and others might fail to notice. These usually come to light when experienced operators are properly trained and motivated to pursue ways of enhancing the job they perform. Operators on key equipment have to understand that eliminating downtime, scrap, and rework, among other wastes, is seen as being just as important as keeping the machine in good operating condition. But such an understanding doesn't always come naturally. Operators have to be trained to look for opportunities and this is especially true of the Lean equipment specialist.

Conducting an Annual Structured Lean Audit

One of the best ways of ensuring an appropriate level of attention is paid to the implementation of Lean is an annual structured audit. Performing this encompasses every aspect of the process from training the workforce, to implemented changes on the shop floor, along with the overall progress achieved against an established and approved master plan for Kaizen.

The team selected to perform the audit should be headed by the Lean coordinator and include representatives from every major support function. Out of the audit should come a recognized winner for the area of the factory that has made the most influential impact in the overall application of Lean. At the conclusion of the audit, a formal report should be prepared by the Lean coordinator and presented to the plant manager, along with posting a summary of the audit for everyone in the factory to see and review.

There are numerous ways for constructing a worksheet for the audit, which should be left to the discretion of the Lean coordinator, but should include:

■ A measure of activities against the assigned plan for Kaizen
■ A measure of costs against the assigned budget for Kaizen
■ A measure of the effectiveness of the maintenance function's participation (help and assistance) in Kaizen-related activity

To perform the audit in the best manner, the participants of the auditing group would normally be broken into two teams: one for the factory floor and the other for the office area. During the audit, every prescribed functional area would be audited. This means areas such as receiving/inspection and shipping, as well as the areas where production work is performed. It should also include office functions.

There are three basic things the area supervisors and department managers are asked during the audit:

1. What Kaizen activity has been performed over the last 12 months?
2. Who was involved in that activity?
3. What specific improvements were made?

The area supervisor or department manager is then asked to show the auditing group the improvements made. Lastly, an effort is made to determine the cost involved and to make note if anything was considered and tabled due to cost restrictions, or the perception that the cost for the effort would simply be too great to absorb. There is a special reason for this. Even under the most conscientious efforts in support of Lean, managers and supervisors can sometimes convince themselves that the cost of an improvement simply can't be justified. Spelling such things out in the audit report allows management the chance to evaluate and decide if the delay was truly warranted or whether instructions should be given to proceed with the change to completion.

Sharing the Results of the Audit with the Workforce

Just as important as performing a structured audit is to share the results with the workforce. This can be done in a number of ways, but one of the best times to hold a general communication session with the entire workforce is after an annual audit of Lean has been conducted. Doing so says that the implementation of Lean is important enough to bring everyone together to speak about results, along with any course correction that may be required and the need for continuing support from everyone in advancing Lean. At this communication session, a selected winner for the most influential impact with Lean should be announced and duly recognized.

The criteria for establishing the winner should center on the three basic areas addressed during the audit, which again is what Kaizen activity was conducted, who was involved, and what specific improvements were made. In most cases, the winner will be easy to define. However, if there is any question regarding one area versus another, "cowinners" should be selected and recognized accordingly.

An annual Lean audit is also a good time for the senior management team to assemble after the audit has been conducted, in order to address progress and any changes or revisions that should be made to the master plan and timetable for overall implementation.

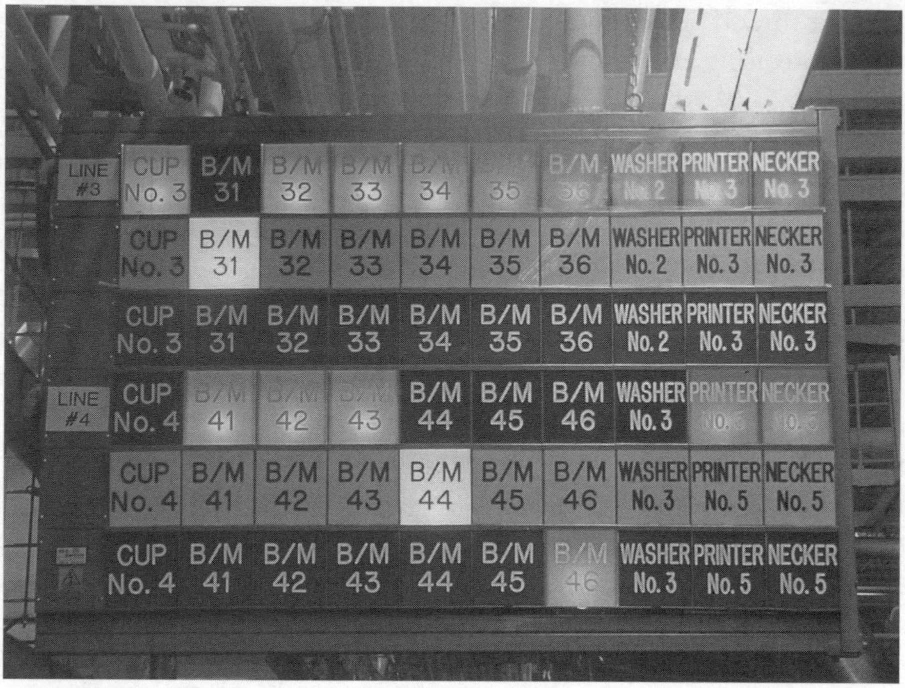

Exhibit 5.1 Electronic final assembly Andon board.

Building in Essential Visual Controls

The subject of visual controls calls for a special word, inasmuch as this is both basic and essential to any good Lean Manufacturing effort. A visual control can be something as simple as pointing out the need to stop and look before entering a high-traffic zone, to something as complex as the example shown in Exhibit 5.1.

In this particular example, each established workstation on a series of final assembly lines is shown on an overhead electronic board, strategically located for visibility. If a workstation light is off, it indicates the station isn't being used for the particular product being assembled. A green light is activated by the supervisor for each workstation being used once a production run is under way. If the operator has something that needs attention he pushes the yellow button on a small console, indicating help is needed. If the operator should be required to stop production because of a quality issue, for example, he pushes the red button on the console, which indicates the line is down and that immediate assistance is required. The yellow caution light can be used for any number of reasons, including the need for special direction or assistance, the potential of running out of needed material, and so forth. But a red light is only activated in the case of completely stopping a workstation, which is turn shuts the entire assembly line down until the problem is addressed and resolved.

With a quick glance what can be seen in Exhibit 5.1 is that Line #3 is utilizing 10 of 30 workstations for the product being assembled and that all the required stations are up and running. Station B/M 31, however, is displaying a yellow light which indicates production hasn't stopped but help is needed. For Line #4, 7 work stations are being utilized for the product being assembled. Station B/M 44 is displaying a yellow light, indicating help is needed. But station B/M46 is displaying a red light, indicating the line is stopped and will remain so until the problem can be resolved.

Most visual controls (Andon) are typically far simpler and less complex; such as shown in Exhibit 5.2. The important thing initially is to get visual controls in place as quickly as possible and set aside any worry about how professional they may look. They can always be dressed up at a later date, through sustaining Kaizen activity.

The goal, of course, is to strive to keep manufacturing running like a highly tuned, well-oiled machine, and good visual controls are essential in striving to meet that objective. But a word of caution should be extended. Never begin the insertion of visual controls unless there is a solid commitment to keep them fully intact and updated as required. Doing so requires assigning responsibility for both the development and upkeep of visual controls, along with appropriate follow-up and audit procedures to assure compliance.

Simple Visual Controls

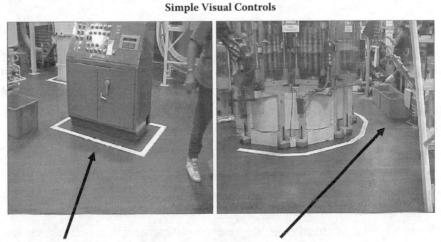

White tape to mark off boundaries.   Red bins = Questionable parts

Exhibit 5.2   Example of simple visual controls.

Constructing a Master Kaizen Plan

Getting the best out of Kaizen and striving to make it a formidable competitive weapon consists of thoughtfully planning its use and the overall path of implementation. Figure 1.1 outlined in Chapter 1 can be extremely helpful in designing a master Kaizen plan. But regardless, the following items should be addressed and preferably constructed in the plan in the order provided:

1. Initial Lean awareness training for all managers, supervisors, and production engineers.
2. Special (extensive) training for the production engineering function in SMED, standard work, Poka-Yoke, and TPM.
3. Inclusion of an effort to determine which equipment fits a "key production equipment" definition.
4. Selection of the area of the factory where the plant's first high-impact Kaizen event will center efforts. The objective of this event should be to establish an area that incorporates most of the aspects of a fully inserted Lean approach to production. In most cases a complete rearrangement of equipment will be necessary in order to effectively demonstrate how Lean should work and function. Visual controls should abound and one-piece flow and pull-production techniques should be applied, along with other actions, such as performing at least one setup reduction (SMED) and mistake-proofing (Poka-Yoke) project. This in turn means a very flexible and responsive maintenance support effort is required during the event.
5. A progressive series of steps involving the use of training and implementation Kaizen events, in order to expose the entire workforce to Lean Manufacturing over a specified period of time (ideally no more than a 12-month period) and to begin the implementation of Lean practices on the shop floor. After the completion of item 5, the next area of focus for the plan should be to apply what is outlined for items 4 and 5 to the office arena. During the process of performing analysis of items 4, 5, and 6, the plan should ideally include production engineering working simultaneously to "Lean engineer" the plant's key production equipment.
6. The introduction of a formal WRAP initiative, followed by a series of progressive steps involving the use of sustaining Kaizen activity and problem resolution Kaizen to further enhance progress and keep a continuing effort going in making a full and complete shift to Lean Manufacturing.

7. At this point in the plan, another high-impact Kaizen event should be considered, centering on making a complete new layout of the plant, in order to take advantage of the space reductions gained and to further enhance pull-production techniques and Lean practices.

Other considerations for the plan should include:

1. A vendor certification process
2. Where applicable, the use of a Lean consultant to assist with certain aspects of implementation, where it is clearly obvious special training or expertise is warranted in order to successfully carry out the plan
3. Deciding specifically where Kanban should be applied

Vendor Certification

The intent of what follows is not to provide a complete and all encompassing overview of a vendor certification program. That would require a book unto itself. The objective is to point out various things that should be considered in the pursuit of vendor certification, which relate to getting the best out of Kaizen and aggressively advancing a Lean initiative.

Typical vendor certification involves such things as building an understanding and agreement on the type of delivery required, the expected quality of the goods received, the class of ISO certification, any subcontracting of the services involved, and so on. Most important to a Lean initiative, however, is for the vendor to reach and maintain a quality status where parts can be delivered directly to the point of use, in the exact quantities specified. In most cases this will not happen with initial certification practices and requires a special effort on the part of both the customer and the supplier.

Once achieved, it would serve to establish the vendor as a class "C" Lean supplier (or some similar form of defined recognition for the effort involved). But in order to achieve the supreme vendor classification (class "A") the supplier would have to demonstrate that they have endorsed and implemented Lean Manufacturing techniques in their factory and must follow this by demonstrating results in the form of a freeze on prices for a fixed period of time and/or a significant improvement in delivery lead time. In turn, the customer usually promises to assist the vendor with training and with certain aspects of Lean implementation. All in all this serves to establish a partnering arrangement that can benefit both parties over the course of time.

Very important to this matter is deciding how far various stipulations for the Lean ratings involved should go and the precise commitment the company is willing to make, with respect to a long-term purchasing agreement. It is a somewhat delicate balancing act, inasmuch as tomorrow can always bring a completely new vendor on the scene that offers as much or more than stipulated, for less cost.

In the late 1970s and early 1980s Toyota set out not only to establish special requirements for vendor certification in support of Lean, but it went so far as to provide land for select suppliers to build factories or supply depots in close proximity to Toyota's main operation. Long-term purchasing agreements with select suppliers proved to benefit both parties initially, but later sometimes worked to a disadvantage when new and more advanced suppliers, operationally, came on the scene. Toyota still follows the practice, but has taken general vendor certification to a much higher level and requires all its suppliers meet quality and delivery standards that fully support the Toyota Production System. Anything less, regardless of price, simply isn't deemed acceptable.

Vendor certification procedures should include three basic stipulations, for any and all suppliers interested in a long-term continuing relationship:

1. That the supplier will diligently work to incorporate Lean Manufacturing practices
2. That the supplier will match, at a minimum, the price of any other viable (proven and acceptable) vendor who may surface later
3. That the supplier willingly takes full responsibility to correct any delivery or quality-related issues that may arise, at their own expense

In turn the company would agree to a fixed level of volume, which should not surpass 60% of a company's total sales volume, unless a very conscious decision is made to do so. Tying one's long-term contractual obligations to 100% of sales volume leaves absolutely no room to maneuver should a vendor with a better price and delivery capability emerge, which even under the best of partnering arrangements can happen.

Ten Commandments of a Fully Supportive Maintenance Function

Maintenance is the single support function in a factory that holds the power to make or break a Lean initiative. Therefore, making certain the maintenance function is both fully capable and properly aligned is absolutely critical to a good Lean Manufacturing effort and especially one that strongly utilizes Kaizen as the tool for advancing the process.

There are ten commandments that apply to a highly supportive Lean-oriented maintenance function:

1. Maintenance shall report directly to the Lean/Kaizen coordinator.
2. Maintenance shall energetically work to support any and all Kaizen efforts conducted, both formal and informal in nature.
3. Maintenance shall carry an appropriate amount of stock on hand for items typically constructed during Kaizen activity, including material to build special fixtures, transfer racks, Andon devices, shadowboards, and the like, and will be highly reactive to such requests when called upon.
4. Maintenance shall have a representative at each formal Kaizen event and be prepared to make equipment rearrangements and provide other needed maintenance support, at the participating team's direction.
5. Maintenance shall be one of the best trained departments in Lean Manufacturing techniques and shall keep such practices in mind with any work they perform.
6. Maintenance shall lead the incorporation of TPM, keeping appropriate records and seeing a full and complete application is applied throughout the facility.
7. Maintenance shall have no words of despair regarding Lean.
8. Maintenance shall know where to quickly obtain outside support in conducting special equipment rearrangements or facility modifications, as needed.
9. Maintenance shall never be critical of ideas submitted by a participating Kaizen team in support of Lean implementation.
10. Maintenance shall apply good Lean practices to the maintenance function itself.

Two typical concerns that have been voiced when I've brought up the ten commandments noted to various manufacturing firms is a feeling that (1) having maintenance report directly to the Lean coordinator and (2) ensuring maintenance is one of the best-trained functions in Lean, could serve to lessen attention on the more typical aspects of maintenance (general facilities upkeep and the like). This simply isn't the case. When a serious condition develops that needs the attention of maintenance, everything takes a backseat to keeping the plant up and running, including Kaizen activity. On the other hand, as more and more work is done in response to a good Lean/Kaizen initiative, fewer and fewer emergencies that require immediate maintenance will tend to develop. It's all a matter of applying good common sense to the effort. However, without clearly aligning maintenance to actively assist in making Lean a full reality, the odds of satisfactory progress are slim, at best.

There are typically cases where the maintenance function has the need for an increase in staff in order to fully support a good Lean effort. This shouldn't be significant, but could include adding one or two additional maintenance employees for a period of time (a year to eighteen months).

Briefly Addressing the Cost and Payback of Lean Again

As pointed out earlier, fully incorporating Lean doesn't come without a reasonably substantial cost. This involves such things as hiring a full-time Lean/Kaizen coordinator, assigning a potential assistant to the coordinator, requiring a possible increase in the production engineering and maintenance staffs, along with a substantially increased level of employee training. But the price paid can have an excellent payback if done correctly.

A company should not be sold on hype that Lean is essentially free, except for some minor equipment rearrangement, painting some new lines on the floor, and doing some basic workplace organization along with some decent housekeeping. That, in reality, is nothing more than window dressing and buys a firm little in the overall scheme of establishing and maintaining a solid competitive position for the future.

Fully incorporating Lean requires a willingness to spend the time, energy, and (where needed) the money to make it happen. But there's an extremely positive side of the equation. Under a soundly constructed and aggressively run progressive Kaizen approach, the payback will start to become clearly evident after a few months, when on-hand inventory levels begin to fall drastically, along with scrap and rework costs and other common wastes that have plagued an operation for years. In addition, it will start to become extremely apparent that manufacturing lead-time can be reduced and that overall product quality has improved.

There will also be signs that space is being cleared, which at some point can be used to rearrange factory equipment and production processing in order to make room for additional products or added business. There will additionally be evidence of an increase in the flexibility of the operation and the overall efficiency of operators, leading in turn to substantial productivity improvements, reductions in downtime, and a steadily declining amount of indirect labor required. Within 12 to 18 months, the payback can start to be measured in terms of multiple thousands of dollars (if not hundreds of thousands of dollars and possibly more).

Ten Most Important Factors to Keep in Mind

The step charts outlined in Chapter 2 and further illustrated in Figure 2.4 provide a roadmap for implementing Lean from start to finish, the finish, of course, related to having changed the system of production and laid the proper foundation for continuous improvement, which is never-ending. Many factories in the United States, however, aren't at the point of starting from scratch. They are at various stages of implementation. The step charts therefore have to be used in adjusting strategy and tactics in order to achieve a full and effective change to the system of production.

Figures 5.2 and 5.3 provide a first and second phase illustration of the 10 most important factors to keep in mind, regardless of a factory's starting point with Lean.

Phase One: Setting the Stage

FACTOR ONE: Learning to Trust the Process

Something that can't be overlooked as extremely important is coming to fully trust the process. It's a certainty that there will be ups and downs, times when outside pressures serve to distract from a focus on Lean and when various conditions appear to tie an operation's hands in advancing the process. Although these have to be addressed and resolved as they arise, the ultimate leader in charge (most often the plant manager) has to ensure they do not become serious stumbling blocks to the full implementation of Lean and the use of an enhanced Kaizen process as the primary tool in making it

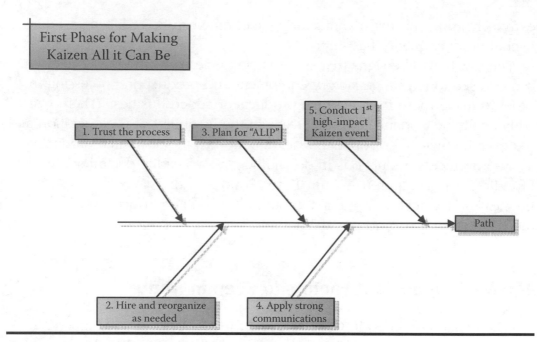

Figure 5.2   First phase of 10-step roadmap.

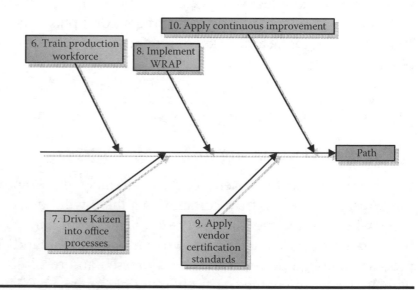

Figure 5.3   Second phase of 10-step roadmap.

happen, he or she has to have a solid trust in the process and a determination to see it through to completion.

One of the best ways of building trust and confidence is to seek out and visit a factory that has taken Lean as close as possible to the ultimate level of accomplishment. What one will always find is an extreme level of pride in the achievements made and more often than not a very strong interest in showing others the accomplishments that have served to make a significant difference. In addition to helping build confidence in the process, there will always be good ideas to take back to the factory and pursue, which can fit in the overall scheme of thoroughly introducing Lean.

Lean Manufacturing always works—regardless of the type of operation, the products produced, and equipment and facilities involved—if a solid set of good introductory principles are followed and a clear path to implementation is established. However, getting a portion of the way there and becoming bogged down can actually do more harm than good. How? By inadvertently creating a hybrid system of production that utilizes a combination of both batch and Lean procedures. This will almost always guarantee a reasonably high level of workforce confusion, if not some seriously strong frustration.

A good reference to keep in mind is the "F" alliance outlined in Chapter 2 (Figure 2.2), which points out the four elements involved in penetrating to the core of Lean implementation. As noted, there must be focus on the mission, faith in the process, the fortitude to fight opposition, and a dedication to seeing it through to a full and successful finish. The overriding factor involved is to strive to keep Lean thinking in mind as one goes about performing normal duties and responsibilities. For a few of the more important players in the process, the following words of wisdom are provided.

For the Plant Manager

Stay positive about Lean! There will be times when conditions will apply stress, whether seeking Lean implementation or not. It's the nature of the beast of manufacturing. But always keep in mind that although stressful feelings can do no harm, how you react to them can. Make a point to occasionally tour the factory with Lean in mind. Talk to operators, floor supervisors, and others about the process and strive to find out what can be done to help advance Lean plantwide. Keep faith that the process will steer things on the correct path, even when it isn't easy to see light at the end of the tunnel. But never be fully satisfied with progress. Strive to keep the attention level high for a march to manufacturing excellence and most important, don't allow Lean to become bogged down by the problems and issues inherent in an old, inefficient, and obsolete system of production. As the old saying goes, "Keep your hand on the throttle and your eye on the rail."

For the Lean Coordinator

Make certain that some level of Kaizen activity is going on in the factory each and every week. Furthermore, make a quick tour at least twice weekly to see for yourself if changes made in support of Lean remain fully intact. Where any slippage takes place, talk to the floor supervisor and strive to get it corrected as quickly as possible. Where there is any lack of support in getting things back in order, don't hesitate to bring it to the plant manager's attention and request his support in fully correcting the situation. Keep a close eye on the Kaizen master plan and where it becomes evident that progress is falling short of that plan, immediately address the matter with the powers above and work to steer things back on course.

For the Production Manager

As the production manager, it's important to keep Lean in mind when addressing problems and production issues that arise. Instead of the old way of addressing problems, look for ways where SMED, Poka-Yoke, Workplace Organization, and the like can be used to correct the situation. For example, would holding a special problem resolution event be helpful in addressing and resolving a nagging production issue? Work to keep the attention level on Lean high on your list of priorities and use every opportunity you can in encouraging the use of Lean practices.

For the Shop Floor Supervisor

Call on the production manager, the Lean coordinator, the production engineer assigned to the area, and others as needed, in helping your people (those that report to you) implement change in keeping with good Lean principles. Be the first to request a training and implementation Kaizen event in your area of responsibility, as soon as it can be scheduled. Expressing an interest in doing so will help in establishing priorities for a Lean effort, because having a supervisor who calls for the introduction of the process always weighs heavily on when and where an event will be scheduled and conducted. If and when a WRAP initiative is undertaken by the company, strive to be one of the first to work with operators in making acceptable improvements. Set goals for operator participation in WRAP and work to see they are successfully achieved. Remember, your principal objective should be to motivate participation and help operators achieve improvements.

For the Production Engineer

Make a practice of performing your assigned sustaining duties by always keeping Lean in mind. This will help in constructing operating methods and routing procedures that do not inadvertently become a hindrance to Lean implementation. Also keep that in mind when planning and procuring new equipment. Construct a Lean equipment checklist (as spelled out in Chapter 4, under "Modifying the Rules for the Purchase of New Equipment") and see that this is duly applied. Lend as much assistance as possible in supporting a WRAP initiative and seek out areas where your help and influence can be utilized in making it a full success. Remain aware that "Lean engineering" key equipment is essential to a good Lean Manufacturing effort and apply all the influence you can in establishing this as a high priority for production engineering.

FACTOR TWO: Assigning Appropriate Talent

Regardless of the best intentions, the job of implementing Lean cannot be successfully carried out without hiring or assigning the appropriate talent to carry it forward. In most manufacturing firms that have been into conventional manufacturing for years, the type of talent needed can indeed be developed with appropriate training and guidance. However, doing this can also be a time-consuming process and it is good to remember that time really isn't on your side. The best approach is a combination of new and experienced talent, along with existing personnel who can be properly trained, guided, and motivated as needed. The most obvious position for the selection of new talent is the person holding the role of Lean coordinator. However, there are other positions that should not be overlooked in striving for a successful Lean venture, which include:

1. The factory's production manager
2. The maintenance manager or supervisor
3. The production engineering function, as a whole

A Bit More about the Production Manager Position

The actual title for this position varies from company to company, but the production manager is the individual commonly in charge of meeting the factory's established production schedules and directing the activities of the production workforce. In keeping with this, all shop floor supervisors normally report to the production manager. It is a highly important job and requires strong leadership ability, one that a company cannot afford to assume will be supportive of Lean and doing all the position can to advance implementation. It's important to remember that most production managers, who have long worked under a conventional system of production, have batch manufacturing deeply embedded in their thinking. Even after training, some of them find it extremely difficult to buy fully into the process.

RELATED EXPERIENCE: I was hired by a family-owned business to help lead the introduction of Lean in a factory that had been into batch manufacturing for over 40 years. Part of the task was to provide advice on properly organizing for the effort. We started with a high-impact event in the final assembly and shipping area of the factory. The first warning sign came when the production manager proceeded to convince senior management that because a number of his key supervisors were required to attend the event, it was vital that he didn't participate and "look after things" to see that production schedules were fully achieved.

I immediately approached the owner of the business to strive to convince him it was vital to the success of the event to have the production manager as a participant and asked if there wasn't some way to free him up for the effort. I went as far as committing to work at reshuffling the training into a six-hour daily schedule for the first week, in order to free the production manager up for the first two hours of the day, so he could "look after things," as he put it.

Senior management went along with the suggestion and everything proceeded as planned. As the event progressed, it became increasingly apparent the production manager had no real interest in being there and held a tremendous amount of doubt about the process being outlined. He grew increasingly late in getting back to class after scheduled breaks and lunch periods. On the fourth day of the event he failed to return to class following the afternoon break. When I approached him late that afternoon he told me an "emergency" had arisen that required his attention. When I went as far as inquiring what the emergency was, he became extremely defensive and unleashed his frustrations in a storm of anger about the entire event and the impact it was having on his "doing his job."

After he cooled down and not so sincerely apologized for the outburst, I told him the important thing was for him to decide if he could live with the kind of change we'd been discussing, because I had little doubt management fully intended to take the process forward. He gave me a very sober expression and said, "Frankly, I'm not sure the company intends to take it as far as you think."

Being a relatively headstrong individual myself, my first reaction was to take the matter to senior management and see if they couldn't somehow put the production manager on the right path. I decided, however, that taking that step could create more distraction than striving to live with his frustrations and allowing the power of the process to convince him, knowing it was going to require some added effort on my part to deal with the situation.

I was correct about the added effort, but the changes the team set out to accomplish were made. After the event concluded and the march for further implementation of the process began in earnest, the production manager became a persistent stumbling block, even as it became increasingly evident the changes being made were clearly to the benefit of the factory. Four months into the process, company management decided to bite the bullet and removed him from his job, giving him a role in scheduling. Two months later the man left to take a position as production manager for a factory that used a conventional batch manufacturing approach and had apparently indicated it had no immediate plans to change its system of production.

I wish the story had a happier ending, that I had somehow been able to convince him with just the right words and he'd gone on to make a remarkable turnaround. The truth, however, is that it seldom happens when it comes to managers who are solidly sold on the belief there isn't a better way than what they've been successful with (in their own mind) for years on end. They are not bad employees. In fact, most are extremely conscientious people who are simply misguided. They fail to see that the old way of doing things doesn't fit with what it takes to stay competitive in the future. This feeling is usually elevated by the fact that they've been praised and patted on the back for years. Accordingly, frustration and resentment can often start to set in when they suddenly hear what they've been doing simply doesn't meet the needs of the future.

Strongly embedded paradigms and their potential influence are two of the chief unknowns attached to getting Lean off to a sound start, but it's something every company will be forced to deal with eventually. Had management taken the time to have a serious discussion with the production manager about where it intended to take Lean and the role he needed to assume in fully supporting the effort, it's likely it would have been obvious he really wasn't the man for the job. A company can always strive to bring the person along with proper training and motivation, but in some cases this will only serve to delay the inevitable and can potentially create a seriously negative influence that really isn't necessary.

I firmly believe every effort should be made to avoid replacing someone who has given an operation many years of service. On the other hand, there is a time and place to take strong organizational action and this is especially true when venturing into Lean. A company has to have a group of managers and key players who express energetic support for the effort, because the change conducted centers on a massive restructuring of a company's way of doing business and comes to affect almost everything and everyone.

Bottom line, the worst assumption a company can make is holding the belief that every important player involved will meet the need adjustment required for Lean, given an understanding of the task. Experience has shown this simply isn't the case.

As far as any rule goes that applies to this matter, it would be: obtain a "feel" for how key personnel will indeed adjust to support the effort. Take nothing for granted and don't hesitate to make a change organizationally if it's evident there's a key player involved who is potentially going to pose some major difficulty with the process. This isn't always an easy or comfortable thing to do, but it's something that can't be overlooked in importance.

FACTOR THREE: Doing the Planning Required to Put ALIP in Place

ALIP again is an Advanced Lean Implementation Process consisting of three working elements: Progressive Kaizen, the four guiding principles of Waste-Free Manufacturing, and the insertion of WRAP (Waste Reduction Activity Process). These work together to establish the fastest and most effective application of Lean Manufacturing in a facility. Can one get there without applying one or more of them? Perhaps, but the journey will be longer and the path more problematic. There are situations where Lean has been successfully introduced without using the process outlined for ALIP, but in most cases it came about due to some very unique circumstances:

1. The company had an adequate, experienced engineering staff and others who were assigned the responsibility of applying Lean principles (i.e., laying out equipment and flow, upgrading work stations, etc.) independent of any input of the operators involved or other influences. In most cases, however, this was done in conjunction with starting up a new factory, quite often on foreign soil.
2. The company hired a plant manager who was highly skilled in implementing Lean and provided the position with the authority to make the kind of change needed with no major restrictions. The basic job outlined was simply: make it happen.

Most factories with an interest in Lean do not have such conditions on which to rely. Lean becomes something they must implement to the best extent possible with what they have to work with, and they have to do so while operating under the parameters of the old system of production. This requires good planning, a very good strategy for implementation, and bringing along the workforce as the effort proceeds.

FACTOR FOUR: Applying Strong and Effective Communications

This step starts with a well-constructed message to the entire workforce spelling out the change needed and how it will be pursued. Employees should know the change will come to oppose the way they've been taught to do things in the past, but for the most part will serve to make their jobs easier and more productive.

Communications relative to the process of change have to be continued on a frequent basis, at least initially. To a large extent it is essentially a "selling" job on the front end. As time goes by, it turns into a matter of providing more specific direction and encouraging steady participation. But looking at it in any manner, good persistent communications are essential to making Kaizen all it can be.

FACTOR FIVE: Demonstrating the Process

The first high-impact Kaizen event sets the stage for where the plant is headed and provides the workforce with a snapshot of how the plant will come to look and operate. I sincerely believe it is a demonstration that must absolutely occur if any solid progress with Lean is expected to be made. If done right it sets the perfect example, an example anyone can look at to get a good idea of what Lean is all about and what's coming down the road. This, of course, requires ensuring that the changes made remain intact and are built upon as time goes by.

Phase Two: Completing the Mission

FACTOR SIX: Training the Production Workforce

At this point things should be in place to begin a seriously active training program for the production workforce. This should include training and implementation, along with problem resolution, and sustaining Kaizen events and associated activities, and should be carried on without any serious disruptions until the entire production workforce has had hands-on exposure.

FACTOR SEVEN: Driving Good Lean Practices into Office Processes

This can begin as soon as the Lean coordinator has help in both teaching and coordinating overall activities. The Lean coordinator will have his hands full with the manufacturing shop floor. Any effort to drive Kaizen into the office arena likely will not happen until the Lean coordinator has a qualified assistant who holds "Office Kaizen" or "Business Process Kaizen" as a chief responsibility.

FACTOR EIGHT: Advancing Improvements at an Individual Job Level

As noted, WRAP establishes an incentive for making Kaizen-related improvements at an individual job level. There are some who would argue an incentive should not be offered for something employees should normally be doing as a matter of practice. To that I can only say there is nothing "normal" about Lean. It's a process that calls for doing things very differently and most employees will not extend themselves to see that change is made without a serious motivator. If management truly feels it can be that motivator by simply instructing employees to participate and by the nature of doing so can get the kind of results needed without offering a monetary award, they should forget about WRAP and move on. However, under normal circumstances WRAP or something similar is needed.

FACTOR NINE: Applying Lean-Oriented Vendor Certification Standards

Vendor certification standards in support of Lean were briefly addressed earlier in this chapter. Again, the idea is to develop and apply things, in support of Lean, to a company's standard vendor certification process. If one isn't in place, start one in support of Lean. There isn't much that is very complicated about it. Work to have select suppliers certified to deliver parts and components directly to point of use, without going through a receiving and inspection process. As time goes by, work to have more and more vendors obtain this capability and focus on eliminating the cost of incoming, receiving, and inspection. This may never be 100% realized, but the cost associated with receiving and inspection (which is classified under most circumstances as an "essential nonvalue waste") can be driven down substantially. Certified Lean suppliers would also work to deliver parts in special containers designed to reduce handling and potential damage, as needed, among other things associated with good Lean practices.

FACTOR TEN: Focusing on Continuous Improvement

Continuous improvement starts the moment the first Lean application is put in place in a factory, and never ends. However, it is after an operation has made a substantial shift to Lean across the entire factory and in the office arena that a renewed effort should be placed on making aggressive, ongoing improvements. That is precisely where Toyota stands today, making strong aggressive improvements to a system of production that would be the envy of most manufacturing operations.

A Final Word

After 25 years in manufacturing and having the privilege of working with countless manufacturing firms in the United States and around much of the world, I have nothing but the utmost respect for the men and women who devote the time, energy, and dedication to providing something of value that serves the needs of others. They are truly pioneers.

I say that because although untold production processes and concepts have been extended their way over the years, starting on a dirt floor in Henry Ford's first attempt to bring an affordable automobile to America's citizens, it has been the men and women of manufacturing who have put those ideas into action and made them a reality. Manufacturing leaders owe these men and women the best of their knowledge and ability because America truly needs a strong manufacturing base and the power and influence it can extend in keeping our country a great and vibrant economy.

Lean Manufacturing offers what could be the last good chance at achieving and maintaining that end. Thus, every effort possible should be extended in making it happen in an effective manner. Following the correct path, it's indeed something that's achievable in any factory.

Something stated in the opening should be repeated in conclusion. "Kaizen isn't limited to the single purpose of making small continuous improvements. Used in the correct manner, it can serve as the chief mechanism in fully inserting Lean Manufacturing throughout an entire business enterprise."

Key Summary Points

■ Although it isn't something most businesses would desire, the work performed on standard shelf-item equipment could be outsourced if the need arose. That simply isn't the case with Class II and Class III production equipment. These are essentially the lifeline of the business and if and when they fail to operate as intended—or fail to operate at all—a business is in serious jeopardy.

- The LE (Lean Equipment) specialist effectively owns the performance of the equipment and holds an above-average responsibility to ensure what comes off that equipment fits the needs of the next user in the process.
- One of the best ways of ensuring an appropriate level of attention on the implementation of Lean is an annual structured audit. Performing this encompasses every aspect of the Lean process, from training the workforce to implemented changes on the shop floor, along with the overall progress achieved against an established and approved master plan.
- Most important to a Lean initiative is for the vendor to reach and maintain a quality status where parts can be delivered directly to the point of use, in the exact quantities specified. In most cases this will not happen with initial certification practices and requires a special effort on the part of both the customer and the supplier.
- Maintenance is the single support function in a factory that holds the power to make or break a Lean initiative. Therefore, making certain the maintenance function is both fully capable and properly aligned is absolutely critical to a good Lean Manufacturing effort, especially one that strongly utilizes Kaizen as the tool for advancing the process.
- The step charts outlined in Chapter 2 and further illustrated in Figure 2.4 provide a roadmap for implementing Lean from start to finish. Many factories in the United States, however, aren't at the point of starting from scratch. They are at various stages of implementation. The step charts therefore have to be used in adjusting strategy and tactics in order to achieve a full and effective change to the system of production. Figures 5.2 and 5.3 provide a first and second phase illustration of the 10 most important factors to keep in mind, regardless of a factory's starting point with Lean.

Appendix A:
Recommended Reading

Key Reference Material

Fast Track to Waste-Free Manufacturing, John W. Davis, Productivity Press, 1999.

Lean Manufacturing; Implementation Strategies that Work, John W. Davis, Industrial Press, 2009.

Kanban – Just-In-Time at Toyota, edited by Japan Management Association, translated by David J. Lu.

Poka-Yoke: Improving Quality by Preventing Defects, edited by *NKS/Factory Magazine*, overview by Hiroyuki Hirano [over 240 illustrated examples].

Zero Quality Control: Source Inspection and the Poka-Yoke System, Shigeo Shingo.

Quick Changeover for Operators: The SMED System, adapted from Shigeo Shingo's "A Revolution in Manufacturing," Productivity Press.

Kaizen and the Art of Creative Thinking (adapted from Shigeo Shingo's Idea, Wo Nigasuna), Tracy S. Epley.

Toyota Production System, Beyond Large Scale Production, Taiichi Ohno.

The Idea Generator: Quick and Easy Kaizen, Bunji Tozawa and Norman Bodek.

Going Lean: How the Best Companies Apply Lean Manufacturing, Stephen A. Ruffa.

Appendix B: How to Obtain Direct Assistance with the Process

For those feeling direct assistance is needed with the process outlined, World Competition Consultants (WCC) has a fully qualified resource base capable of meeting a factory's initial counseling and training needs. The services include:

- *Plant Management Overview:* A one-day working session designed to provide plant management and staff with an overview of the outlined process and how it can most effectively be applied, considering the current status of Lean within the company or factory involved. This service is provided free of charge with the exception of travel-related expenses.
- *Lean Coordinator Training:* A two-day training seminar for the Lean coordinator or a number of company or corporate Lean coordinators, designed to provide direction in the use of ALIP (Advanced Lean Implementation Process) and explain the logic and purpose of "Progressive Kaizen." Included are how to go about developing a master implementation plan, and when and how to both use and perform the various types of Kaizen outlined.
- *Production Engineering and Shop Floor Supervisor Training:* A three-day training session designed to provide production engineers and shop floor supervisors with knowledge in Progressive Kaizen, along with the skills needed in fully engineering a plant's key production equipment. Included is an advanced level of training in SMED, Poka-Yoke, and TPM, along with an in-depth explanation of the four guiding principles of Waste-Free Manufacturing.

■ *High-Impact Kaizen Event:* A one- to two-week extensive Kaizen event aimed at making sweeping change to a given area of the factory and establishing a "showcase" that is fully representative of the type of change that will be required for the entire factory over the course of fully implementing a Lean Manufacturing initiative. Included is a free of charge pre-event visit to work with management in establishing which area of the factory would be the most ideal for the event, which support functions should have representative participants, how maintenance should be prepared to respond, and what the expected results should be.

■ *Training and Implementation Kaizen Event:* A three-day Kaizen event aimed at training a select group of hourly and salaried participants in Lean Manufacturing. Half of the event is classroom training with the other half aimed at applying a portion of the training on the shop floor. Typically the magnitude of change centers on applying workplace organization, which utilizes the 6-Cs (clear, confine, control, clean, communicate, continue).

■ *Problem Resolution and Sustaining Kaizen Events:* A three-day training seminar for factory managers, shop floor supervisors, and select hourly and salaried personnel, in how to go about identifying the root cause of recurring production problems and put a permanent fix in place. In addition, guidance and direction are given in how to use sustaining Kaizen, along with how to conduct an SK event: who should be involved, what the training should focus on, and how to measure results.

To obtain assistance or for further information regarding direct assistance in establishing and running a highly results-oriented ALIP and Progressive Kaizen process as outlined:

Email: wfmassociates@aol.com
Telephone: 1-501-884-4862

Glossary: Definitions of Frequently Used Terms

Advanced Lean Implementation Process (ALIP): An all encompassing improvement concept for fully and effectively implementing Lean Manufacturing, made up of three major components: Progressive Kaizen, Waste-Free Manufacturing, and WRAP (Waste Reduction Activity Process).

Conventional Manufacturing: The system of production that grew out of the Henry Ford era and became the standard for U.S. industry after World War II. Built on the driving fundamentals of batch manufacturing techniques, the system endorses the principles of build and queue production and is supported by performance measurements and information systems that serve to perpetuate the process.

Error-Free Processing: One of the four guiding principles of "Waste-Free Manufacturing," which utilizes Poka-Yoke (mistake proofing) and 5-W (the five "whys") as primary tools for establishing root cause and engineering special devices aimed at source inspection and the correction of potential errors in processing.

High-Impact Kaizen: The first of the four components of Progressive Kaizen, High-Impact Kaizen is a structured activity aimed at training managers, along with a select number of hourly and salaried associates, in Lean Manufacturing principles and techniques. Through a combination of extensive classroom training and the application of hands-on waste reduction activity (Kaizen), participants make sweeping change to a prescribed area of the factory, with the intent of establishing a "showcase" area that is representative of the type of production system intended for implementation across the entire factory.

Insignificant Changeover: One of the four guiding principles of Waste-Free Manufacturing, aimed at reducing setup and changeover to the point of becoming "insignificant" to the decision-making process, involving such things as taking on added business, revamping production schedules to satisfy customer needs, and the like.

Kaizen: A Japanese term for a process aimed at making continuous improvement, by focusing on and eliminating wastes inherent to the manufacturing practices being employed. As outlined in this work, the practice of Kaizen is expanded in functional application to become the chief mechanism for advancing the insertion of Lean Manufacturing.

Kaizen Event: A structured exercise consisting of a select group of managers, supervisors, and hourly and salaried associates aimed at providing classroom training in Lean Manufacturing, along with the opportunity for participants to make hands-on change on the shop floor or the office arena, utilizing the tools of the Toyota Production System.

Lean Manufacturing: A widespread term for a manufacturing approach that utilizes the tools and concepts of the Toyota Production System, along with a variety of other improvement techniques such as value stream mapping and Six Sigma applications, all of which are aimed at eliminating wastes, improving flexibility, and better servicing the customer.

Poka-Yoke: A Japanese term for a process aimed at "error-proofing" production operations, derived from Poka (mistakes) and Yokeru (avoid). As a science, Poka-Yoke is designed to control quality at the point of application. As applied in progressive Kaizen, Poka-Yoke is a major component of error-free processing.

Problem Resolution Kaizen: The third of the four components of Progressive Kaizen. A structured process directed at establishing the root cause of recurring production problems and/or issues, and applying a permanent solution.

Progressive Kaizen: The coupling of all the various "types" of Kaizen outlined, with a uniform approach to their use, along with a clearly established plan of action and specific management initiatives that actively serve to promote and advance the full insertion of Lean Manufacturing.

Production/Sustaining Engineers: College graduates in manufacturing engineering, industrial engineering, or industrial technology, assigned the job of sustaining a factory from the standpoint of equipment design and procurement, shop floor methods, direct labor standards, and other critical production support activities.

SMED: A term meaning Single Minute Exchange of Dies. SMED is aimed at the science of reducing setup and changeover on equipment or processing to single minutes (nine minutes or less) by examining and breaking down the work involved into internal and external components and designing methods that allow much of the preparatory work to be performed prior to the time a setup is conducted. As applied in Progressive Kaizen, SMED is a component of the principle of insignificant changeover.

Sustaining Kaizen: The fourth of the four components of Progressive Kaizen; sustaining Kaizen is a process consisting of both formal and informal waste reduction activity, aimed at enhancing previous waste reduction efforts and providing employees a means of making individual job improvements.

System of Production: The policies, procedures, and practices employed to manufacture parts, components, and finished units, along with the prescribed assistance of various support functions and related business information systems.

Total Productive Maintenance (TPM): TPM is a process designed to extend the typical aspects of preventative maintenance to strongly include operators who work in conjunction with the maintenance department with machine upkeep, performing such tasks as the lubrication of equipment and special equipment inspection, on a planned frequency basis. In practice, TPM is a special partnering of the operator, floor supervision and the maintenance function in assuring a plant's production equipment receives the kind of oversight required to keep it operating at an optimum level of performance.

Toyota Production System (TPS): A system of production principally implemented under the direction of Taiichi Ohno and Shigeo Shingo, aimed at eliminating manufacturing wastes, improving overall flexibility, greatly reducing inventory, and substantially lowering operating costs.

Training and Implementation Kaizen (TI): The second of the four components of progressive Kaizen, training and implementation Kaizen is directed at providing basic training in Lean Manufacturing for the entire workforce over an extended period of time and providing participants the opportunity to apply a portion of what they've learned in a given production area of the factory. Due to the short duration of the event, changes made by the group principally center on "workplace organization."

Uninterrupted Flow: One of the four guiding principles of waste-free manufacturing, pursuant to point-of-use, one-piece flow, Kanban, and other applications that serve to eliminate stoppage and storage points in flow.

Waste-Free Manufacturing: An all-encompassing Lean Manufacturing initiative spelled out in my book, *Fast Track to Waste-Free Manufacturing*, which utilizes the tools of the Toyota Production System, under the guiding influence of the principles: workplace organization, uninterrupted flow, error-free processing, and insignificant changeover.

Waste Reduction Activity Process (WRAP): A management initiative that provides a prescribed incentive (bonus) for Lean-related improvements made at an individual job level. The intent is to keep the attention level of the workforce focused on continuous improvement and the elimination of wastes inherent to the job. Acceptable changes come under the four guiding principles of waste-free manufacturing (workplace organization, uninterrupted flow, insignificant changeover, and error-free processing).

Workplace Organization: One of the four guiding principles of Waste-Free Manufacturing, described as the foundation of all continuous improvement. Workplace organization involves setting up an operator-friendly workplace and utilizing the 6-Cs as the primary tool in structuring change (clear, confine, control, clean, communicate, continue). Upon application, the workplace has extensive visual controls built in that greatly reduce the chance of downtime and production errors, along with better ensuring quality workmanship.

Index

10-step roadmap for Lean implementation, 162
5-S, 102
5-W, application of, 127–128
6-C, 102, 119

A

Accomplishments, deterioration of, 70–74
ALIP (Advanced Lean Implementation Process), 96
    components of, 8–9
    planning for, 169
Andon, 102, 119
    electronic final assembly board, 154
Annual structured audits
    conducting, 152–153
    sharing results of with workforce, 153
Audits
    conducting, 152–153
    sharing results of with workforce, 153

B

Batch manufacturing, comparison with Toyota production system, 11
Bonuses, WRAP initiatives and, 54–56
Budget, 21–22
Business process improvement, 100
Business Process Kaizen, 170. *See also* Office Kaizen

C

Change
    distribution of, 36–38

reversal of, 70–74
training and implementation events and, 123–124
Classroom training, 121–122
Communication
    importance of, 74–76
    strong and effective, 169–170
Company president, Lean-oriented, 45–48
Component tool box, 96–97
Consultants, selection of, 64–65
Continuous flow, 132–133. *See also* Uninterrupted flow
Continuous improvement, 171–172
Conventional manufacturing
    comparison with Toyota production system, 11
    comparison with world class manufacturing, 75–76

D

Developing a schedule, 19–22
Discretionary management initiatives, 99
Displaced operators
    defining, 114
    use of, 116
Disposition zone, 119
Disruptive Kaizen, 14
Do's and don'ts of Kaizen, 16–17, 86–88

E

Effective Kaizen, 13–14
Equipment, modifying the rules for purchase of, 124
Error-free processing, 9–10, 98, 101

office personnel and, 68
Evaluating Kaizen efforts, 17–19
Events, 28–30. *See also* Kaizen events
    applied purpose of, 34
    conducting, 22
    high-impact, 107–121
    problem resolution, 128–132
    training and implementation, 121–125
Expenses, Lean Manufacturing and, 84–86,
    94–95, 160

F

F Alliance, 44–45, 138
Final assembly, placing a pull zone in,
    103–106
Finance's role in Kaizen, 69–70, 84–86
Floor exercises, 130–131

G

Group leaders, 147–148
Guiding principles, 8–10, 67, 119–120

H

High-Impact Kaizen, 23–24, 26–27
High-Impact Kaizen events, 143
    conducting, 107–118
    demonstrating the process, 170
    preparing for, 118
    structure and steps involved in
        conducting, 119–121
    wrap-up and follow-up aspects of,
        118–119

I

Improvements, implementing WRAP
    initiatives for, 135–138
Industrial engineering, 76–77. *See also*
    Production engineering
Insignificant changeover, 9–10, 98–99,
    101–102

K

Kaizen
    allowing accomplishments to deteriorate,
        70–74
    communication of extent and scope of,
        74–76
    designing a master plan, 156–157
    developing a formal schedule for, 19–22,
        33
    do's and don'ts, 16–17, 86–88
    effective *vs.* disruptive, 13–14
    elevating the use and effectiveness of, 33
    establishing a formal budget for, 21–22,
        34
    evaluating and rating efforts, 17–19
    implementation of, 5–6
    labeling as waste reduction, 100–103
    overview of various types of, 23–28
    scope of activity, 142–143
    tactics for getting the best results from,
        56–60
    value of WRAP initiatives in, 53–56
Kaizen events. *See also* Events
    applied purpose of, 34
    conducting, 22, 107–121
    defining, 28–30
    role of production manager in, 32–33
Kaizen initiatives, role of shop floor
    supervisor in, 48–49
Kaizen training, scope of, 23
Key equipment, 146. *See also* Lean
    Equipment (LE) specialists

L

Labor classifications, 52–53
Labor pools, creation of separate, 116–117
Labor unions, 52, 55
    participation of officials in high-impact
        Kaizen events, 114–118
Layoffs, displaced operators and, 115–116
Lean accomplishments, distribution of,
    36–38
Lean consultants, selection of, 64–65
Lean coordinators, 43–44, 69, 164

role of in Problem Resolution Kaizen events, 129–130
Lean Equipment (LE) specialists
apprentice training, 149–150
considerations for using, 146–147
owner-operators as, 145–146
pay structure for, 147–148
percentage of workforce classified as, 148
Lean implementation
assigning the appropriate talent for, 165–168
conducting annual structured audits of, 152–153
step chart for, 50 (*See also* Step charts)
Lean initiatives
avoiding disruptions of, 106–107
cost of, 94–95, 160
focus of, 6–7
making Kaizen-related workforce adjustments, 115–116
payback of, 95–96, 161
placing the first pull zone, 103–106
role of plant managers in, 38–41
ten most important factors for, 161–172
training first-line supervisors, 142
Lean manufacturing
conducting factory's first high-impact event, 107–121
creation of Kaizen labor pool for, 116–117
equipment checklist, 125
error of putting in stand-by mode, 84–86
implementing, 6, 35–36
Lean-oriented company presidents, 45–48
Lean-oriented plant managers, characteristics of, 41–43
Lean/Kaizen coordinator, assignment of, 20–22

M

Maintenance, 16
Lean-oriented function of, 159–160
Maintenance manager, 44, 60
Maintenance personnel, participation of in high-impact events, 111–112
Management audits, Lean accomplishments and, 73–74

Management initiatives, discretionary, 99
Manufacturing
common measurements in, 13
role of Kaizen in, 1–4
Manufacturing engineering, 76–77. *See also* Production engineering
Master plans, constructing, 156–157
Measurements, 13, 78
Misguided pragmatism, 15
Mistake proofing. *See* Poka-Yoke

N

New equipment, modifying the rules for purchase of, 124

O

Objectives
high-impact events, 108
stated, 79–84
weighting of, 49, 56–57
Office function, exclusion of, 67–68
Office Kaizen, 68–69, 100, 170
One piece flow, 102
Outside assistance, 64–67. *See also* Consultants
Owner-operators, 60
as Lean equipment specialists, 145
special consideration of, 52–53

P

Paint line supervisors, 10
Paradigms, 15
Payback, 95–96, 161
Planning, 19–22
constructing a Kaizen master plan, 156–157
Plant managers, 60, 163–164
characteristics of Lean-oriented, 41–43
role of in Lean, 38–41
Point-of use manufacturing techniques, 102
Poka-Yoke, 77–78, 101, 119–120, 129–130
Pragmatism, misguided, 15
Proactive maintenance, 76
Problem Resolution Kaizen, 23, 25, 27

driving the use of, 126–127
Problem Resolution Kaizen events, 144
  conducting, 128–132
  essential tools utilized in, 131–132
Processes, versions of, 88–89
Production associates
  role of in high-impact Kaizen events, 112
  training of, 170
Production engineering, 60, 91, 100, 165
  participation of in high-impact Kaizen
    events, 113
  role of in Kaizen initiatives, 76–79
  stated objectives of, 79, 83–84
  utilization of in Lean initiatives, 37–38
Production managers, 164, 166–168
  participation of in high-impact events,
    108–110
  role of in a Kaizen event, 32–33
  stated objectives of, 79–82
Production problems, addressing with
    problem resolution Kaizen (PRK),
    128–132
Progressive Kaizen, 8–9, 28–30
  component tool box for, 96–100
  purpose and scope of effort, 30
  types of, 23–28, 33
Pull systems of production, 10, 76
Pull zones, 143
  primary placement of, 103–106

R

Rating Kaizen efforts, 17–19
Recommended reading, 175
Related experience examples, 2–3, 31,
    39–40, 49, 51, 66–67, 71, 73–74,
    81–82, 88, 101–102, 105–106,
    109–110, 150–151, 166–167
Roadmaps for Lean implementation, 162
ROI (Return on Investment), 22, 94–96
Root cause, establishing, 126–127

S

Salaried employees
  consideration of in costs of Lean
    initiatives, 94

participation of in high-impact Kaizen
    events, 114
  use of Kaizen by, 68
Shop floor supervisors, 164–165
  participation of in high-impact events,
    111
  role of in Kaizen, 48–49
  stated objectives of, 79, 82–83
SMED (Single Minute Exchange of Dies),
    77–78, 102, 119–120
Space gains, taking advantage of, 98–99
SPC (Statistical Process Control) charts,
    126–127
Standard Work, 77–78, 100
Stated objectives, 79–80
  production engineering, 83–84
  production manager, 80–82
  shop floor supervisor, 82–83
Step charts, 50, 57–59
Supervisor training, 142
Sustaining Kaizen, 23, 25, 28
  components of, 134
  understanding role and scope of, 133–135
Swiss cheese phenomenon, 98–99
System of production, 4

T

Tours
  use of to follow-up high-impact Kaizen
    events, 118–119
  use of to prevent erosion of
    accomplishments, 72–74
Toyota production system
  beginnings of, 2
  comparison with batch manufacturing, 11
  Toyota's transition to, 6
  vendor certification procedures for, 158
TPM (Total Productive Maintenance),
    119–120, 131
Training. *See also* Training and
    Implementation Kaizen
  first-line supervisors, 142
  importance of, 138–139
  production workforce, 170
Training and Implementation Kaizen, 23–24,
    27

Training and Implementation Kaizen events, 121–125
  planned frequency of, 125
Trusting the process, 162–165
Types of Kaizen, 23–28. *See also* specific types

U

Uninterrupted flow, 9, 98, 102, 119, 132–133
Unions, 52, 55
  participation of officials in high-impact Kaizen events, 114–118

V

Vendor certification, 100, 157–158, 171
Visual controls, building in, 154–155

W

Waste reduction, labeling Kaizen activities as, 100–103
Waste-free manufacturing, 8–9, 133
  WRAP initiatives and encouraging, 135–138

Weekly Lean tour, use of to prevent erosion of accomplishments, 72–74
Weighting objectives, 49, 56–57, 81–82
Whys, 127–128
Work measurement, 78
Work-in-process inventory, costs of carrying, 95
Workforce training, 170
Workforce-adjustments, Kaizen related, 115–116
Workplace organization, 9, 97–98, 102, 119, 133
  office personnel and, 68
World class manufacturing, comparison with conventional manufacturing, 75–76
World Competition Consultants, 177–178
WRAP (Waste Reduction Activity Process), 8–9, 61, 99, 101, 143–144, 171
  communicating and tracking results of, 139–141
  hurdles to initiating, 138–139
  initiatives, 53–56
  introduction timeframe, 140
  planning phase for, 135–138
  twofold objectives of, 141
  use of in office functions, 68–69